ALKALOIDS – ALTERNATIVES IN SYNTHESIS, MODIFICATION AND APPLICATION

Edited by **Vasil Georgiev** and **Atanas Pavlov**

Alkaloids – Alternatives in Synthesis, Modification and Application
http://dx.doi.org/10.5772/64120
Edited by Vasil Georgiev and Atanas Pavlov

Contributors

Annett Gefrom, Annette Zeyner, Diah Ratnadewi, Isadore Kanfer, Srinivas Patnala, Alejandro Cruz, Itzia I. Padilla-Martínez, María Esther Bautista-Ramírez, Serbulent Gokhan Beyaz, Ahmet Celiktas, Havva Sayhan

Published by InTech

Janeza Trdine 9, 51000 Rijeka, Croatia

Publishing Process Manager Dajana Pemac
Technical Editor SPi Global
Cover InTech Design team

First published July, 2017
Printed in Croatia
Legal deposit, Croatia: National and University Library in Zagreb

Additional hard copies can be obtained from orders@intechopen.com

Alkaloids – Alternatives in Synthesis, Modification and Application, Edited by Vasil Georgiev and Atanas Pavlov
p. cm.
Print ISBN 978-953-51-3391-9
Online ISBN 978-953-51-3392-6

For EU product safety concerns:
IN TECH d.o.o., Prolaz Marije Krucifikse Kozulić 3, 51000 Rijeka, Croatia, info@intechopen.com or visit our website at intechopen.com.

Meet the editors

Dr. Vasil Georgiev obtained his PhD degree at the Stephan Angeloff Institute of Microbiology, Bulgarian Academy of Sciences, in 2009. He is an expert on bioprocess engineering and secondary metabolite production by plant cells, tissue and organ cultures, functional genomics, metabolite profiling, development, and improvement of plant in vitro production platforms. In 2011, he was awarded by the Bulgarian Academy of Sciences with the "Prof. Marin Drinov" prize for the outstanding research in biological sciences. Dr. Georgiev did postdoctoral research at Florida A&M University, USA (2012–2016). Since 2016, he is an associate professor in the Laboratory of Applied Biotechnology at the Stephan Angeloff Institute of Microbiology, BAS, and also joined the Department of Organic Chemistry at the University of Food Technologies, Plovdiv, Bulgaria.

Prof. Atanas Pavlov completed his PhD degree in biotechnology at the Stephan Angeloff Institute of Microbiology, Bulgarian Academy of Sciences, in 1998. From 2003 to 2005, he did postdoctoral research at the Institute of Food Technologies and Bioprocess Engineering, University of Technologies, Dresden, Germany. In 2005, he joined the Stephan Angeloff Institute of Microbiology, Bulgarian Academy of Sciences, as an associate professor. In 2009, he received his habilitation (DSci) and is, since 2010, a professor at the Laboratory of Applied Biotechnology. Since 2011, he also holds a second professor position at the University of Food Technologies on Plovdiv, Bulgaria. His research is mainly focused on the production of specific bioactive metabolites by plant cell, tissue, and organ cultures.

Contents

Preface

Alkaloids are a large group of natural, semisynthetic, and synthetic compounds which contain basic nitrogen. The great diversity in chemical structures, sources of origin, pharmacological activities, and physiological actions predetermines alkaloids as an important research target for many scientists from all over the world. The broad range in their biological activities and the amenability to chemical modifications rank the alkaloids among the most attractive candidates in new drug discovery researches.

The book *Alkaloids – Alternatives in Synthesis, Modification, and Application* collects several researches by eminent authors, summarizing the recent advance in research. The book encompasses a diverse palette of different topics related to sustainable supply of valuable alkaloids, their chemical modification, and synthesis of analogues with improved pharmacological properties and reduced side effects through the concerns for monitoring and standardization of alkaloid content in raw and fermented natural products. The book includes five chapters, guiding us through nonconventional approaches for alternative bioproduction of alkaloids, the application of modern chemical methods of asymmetric synthesis for production of synthetic and semisynthetic alkaloid derivatives, historical overview of medicinal application of alkaloids as anesthetics and in management of pain, application of alkaloid fingerprint for chemotaxonomy, quality control and standardization of plant material, and, not at last, the importance of monitoring and reducing the alkaloid content in animal feedstuffs.

The first chapter provides a detailed overview of recent advances in application of plant biotechnology for in vitro production of alkaloids. Belonging to the class of secondary metabolites, responsible in cell defense, the alkaloids are accumulated in very low concentrations in plant tissue, and, therefore, a processing of large amount of raw material is necessary to secure their regular supply. The chapter describes the recent tendency for the use of plant cell suspension cultures as alternative source for continuous production of alkaloids with valuable properties. Different factors affecting the alkaloid accumulation and strategies for improving systems productivity yields are discussed, and the limitations slowing-down the commercialization of that technology are pointed.

The second chapter discusses the recent methods and synthetic methodology, applied in production of acyclic and heterocyclic derivatives of ephedra alkaloids. Ephedrine is an alkaloid isolated from *Ephedra* spp., which is used as sympathomimetic drugs to treat cardiac arrest, bronchial asthma, obesity, lowering blood pressure, etc. With the advance in modern chemistry, ephedrine has been broadly used in asymmetric synthesis to produce physiologically active derivatives with improved pharmaceutical properties. The strategies for synthesis of such optically pure compounds are presented, and their potentials as candidates for new drugs are discussed.

Plants accumulating alkaloids have been used to relieve the pain by traditional healers all over the world from time immemorial. Nowadays, the modern medicine still relies on these phytochemicals and their closely related synthetic and semisynthetic analogues for anesthesia and management of pain during many medicinal manipulations. The third chapter of this book is a comprehensive review on the history, mechanism of action and medical application of bioactive alkaloids with anesthetic, pain-relief, and psychoactive actions, and their chemically improved derivatives.

The fourth chapter is focused on chemistry and ethnopharmacology of alkaloids from *Sceletium* plants. Special attention is paid on the importance of alkaloid profiling and chromatography fingerprint for chemotaxonomy and identifying of closely related *Sceletium* species. The advantages of proved and powerful analytical techniques such HPLC, LC-MS/MS, and CZE are discussed in a light of quality control and standardization of alkaloid content in *Sceletium*-based products.

The last chapter of the book presents the results of detailed research on improvement of nutritional values of animal feedstuffs. This chapter comes to remind us that the alkaloids are not always desirable ingredients. In a case of lupine grains, the accumulated alkaloids are anti-nutritional factors causing decrease in farm animals" performance. The research investigates the effectiveness of silaging process by using the lactic acid bacteria to decrease the alkaloid content in lupine grains. The importance of monitoring and control of alkaloids concentrations in final silage are discussed.

By accepting the challenge for compilation of this book, we aimed to raise the curiosity and generate new ideas in broad range of students, scientists, chemical engineers, physicians, and young explorers by enlightening some very interesting and highly perspective areas in alkaloid research. We believe that we greatly succeeded in this goal.

We would like to express our sincere thanks to all contributing authors for their efforts and professional expertise in realization of this book. We hoped that our shared efforts will open new horizons in alkaloid research.

Assoc. Prof. Vasil Georgiev
Laboratory of Applied Biotechnology
Stephan Angeloff Institute of Microbiology, Bulgarian Academy of Sciences
Bulgaria

Department of Organic Chemistry, University of Food Technologies
Bulgaria

Prof. Atanas Pavlov
Department of Analytical Chemistry, University of Food Technologies
Bulgaria

Laboratory of Applied Biotechnology
Stephan Angeloff Institute of Microbiology, Bulgarian Academy of Sciences
Bulgaria

Chapter 1

Alkaloids in Plant Cell Cultures

Diah Ratnadewi

Additional information is available at the end of the chapter

http://dx.doi.org/10.5772/66288

Abstract

Alkaloids are natural substances used largely for human medication. Destruction of natural habitat of various alkaloid-producing plants has led to cultivation. But large areas of cultivation are required due to very low content of the substances, which presents a consequence of high production cost. Cell cultures, undifferentiated or differentiated, have exhibited the capacity to synthesize alkaloids *in vitro*, which may be developed as alternative source of phytochemicals. The discoveries of biosynthetic pathways, sites of stepwise synthesis, and accumulation of alkaloids in plant and cultured cells, also factors involving in or influencing those mechanisms, are all mosaics that build an increasingly obvious picture of the case. Despite enhancement in alkaloids production in *in vitro* culture through many kinds of manipulation, variability in the yield still occurs. Improvements on the basic knowledge of cell mechanisms, including metabolomics, along with genetic engineering are expectable to solve the problems.

Keywords: alkaloids, cell culture, *in vitro* production, accumulating structures

1. Introduction

Many alkaloids are pharmacologically active and have been used since decades as valuable medicines. They are produced by a large variety of organisms such as bacteria, fungi, plants, and animals. People in many parts of the globe used to extract natural substances from plants that actually contain alkaloids to cure them against certain diseases or to maintain their health. Alkaloids are known as having antimicrobial and pharmacological activities such as analgesic, antipyretic, antihypertensive, cardiac antiarrhythmic, antimalarial, and anticancer [1–5]. Nevertheless, only few of those natural alkaloids have been investigated through a proper pharmacological test, while we are being run after by massive deforestation and physical development in many countries. Moreover, natural production of alkaloids is quite vulnerable vis-à-vis the climatic condition and produced in very low level. Therefore, one of the efforts to save any invaluable plants' substances from extinction and at the same time to increase and stabilize the production is by culturing plant cells and tissue *in vitro*. A suitable term for this is cell farming.

Secondary metabolites biosynthesis and accumulation are associated with a variety of cell types in different plants, including epidermis, endodermis, pericycle, phloem parenchyma, phloem sieve elements and companion cells, specialized mesophyll, and laticifers. There have been many evidences that plant cells cultured *in vitro* are capable to synthesize secondary metabolites, then people tried to culture tissues or organs of the plant from which a desirable phytochemical used to be highly accumulated, for instances roots for ginsenoside, seeds of *Trigonella* spp. for diosgenine, or leaves of *Catharanthus roseus* for vinblastine. However, the case is not always true, since some alkaloids are synthesized in one plant organ and then translocated to another one for accumulation [6]. Contrarily, plant tissues isolated from different parts of the naturally producing plant have been reported as being capable to synthesize the substance in *in vitro* cell culture; the cells are either undifferentiated or differentiated. Due to this fact, choosing the type of plant tissue is of critical importance for the success. Once a phytochemical production is established *in vitro*, it is subject to upscale the production, particularly of substances which have high commercial value. Through various ways, that is, chemically, technologically, and genetically, some of them have reached industrial cost-benefit level. Metabolic along with genetic engineering work together or in parallel to improve the cells' yields.

This chapter describes *in vitro* production of alkaloids and factors that may influence the synthesis and storage in cultured cells which were found in some cases different from those discovered in plant. It also reports the existence of specialized structures in *in vitro* cell cultures compared to natural secretory cells that occur in intact plant.

2. Alkaloids in *in vitro* cell cultures

Secondary metabolites are generally synthesized in very low level in plants (only up to 1% of total carbon) as they have specific functions. Mostly, they serve in plant protection against biotic and/or abiotic stresses. The use of biotechnology strengthens the efforts to improve the yield of a desired compound. Cultured plant cells can be considered as factory cells that produce secondary metabolites due to their capability for bio-transformations. Thus, some processes such as hydrogenation, dehydrogenation, isomerization, glycosylation, hydroxylation, and addition of carbon atoms might be performed in plant cell cultures.

The biosynthesis pathways of indole alkaloids and the others have been investigated for many years, but the whole process is not completed yet. Good knowledge on the mechanism of alkaloids biosynthesis in cell cultures and cultural condition will considerably enhance the production of natural alkaloids for medication. In plant, the tissue of synthesis is not always the same as the tissue of accumulation. Certain substance is derived from a precursor that presents in one organ, and then the whole pathway from one intermediate to subsequent ones is performed in different organs/sites, each with its specific enzyme. When this is the case, the substance must be synthesized in differentiated plant where translocation takes place from one plant part to another. That substance might not be found in *in vitro* cells culture. Another case is that alkaloid biosynthesis and storage are in a particular organ or tissue; then performing *in vitro* cultures of a specific organ or tissue is likely to get the desired substance. When biosynthesis and accumulation occur in any cells of a plant, the alkaloid of interest can be produced in the culture [6].

It was suggested that the synthesis and storage of quinoline alkaloids in Cinchona plant is determined by the extent of differentiation. Experiments with hairy roots culture of Cinchona [7, 8] resulted in an increased yield of the alkaloids. In mature plant of *C. roseus*, genes expression for tryptophan decarboxylase (TDC) and strictosidine synthase (STR), both are among the key enzymes for terpenoid indole alkaloids (TIAs) biosynthesis, were detected much higher in roots than in the aerial part of the plant [9]. The research results led to the integration of *Cinchona officinalis* hairy roots with *tdc* and *str* genes originated from *C. roseus*, in an attempt to increase the alkaloid production [10]. Indeed, the transformed hairy roots culture produced higher level of quinoline alkaloids, suggesting that in *Cinchona* spp., roots may be the main site for those two enzymes. Many other researchers reported that undifferentiated cultured cells of Cinchona exhibited quinine alkaloid in control medium [11, 12] as well as in elicitors treated medium [12]. This means that all Cinchona cells can synthesize the alkaloids in some ways.

Similar stories also applied to reserpine biosynthesis in *Rauwolfia serpentina*. Reserpine was found in cells or callus culture of the plant [13, 14] as well as in its shoots culture [15], while naturally people collect reserpine from its roots. Dimeric TIAs, vinblastine and vincristine, have been detected in higher level in new-formed roots and callus-roots cultures of *C. roseus*, compared to that existed in its petiole [16]. Those examples suggest that *Rauvolfia* spp. and *Catharanthus* spp. are capable to synthesize alkaloids regardless of the tissue type as biosynthetic site.

Regenerated shoots of *Papaver bracteatum* contain much higher thebaine/morphine alkaloids than meristemoids or callus. The shoots have laticifer-like cells and tracheids, whereas meristemoids and callus occasionally do not show the presence of laticifer [17]. It was reported also that only greenish callus with tracheary elements produced thebaine and codeine in *P. somniferum* [18]. Ginsenoside is specifically produced in ginseng roots; therefore, *in vitro* root culture is required. Similarly, herbal plants such as *Hypericum perforatum* (St. John's wort), which accumulates hypericins and hyperforins in foliar glands, have not demonstrated the ability to accumulate the phytochemicals in undifferentiated cells [19]. Those cases indicate that cytodifferentiation has correlation with alkaloids synthesis, and most presumably cells culture of those certain plant species cannot produce the corresponding alkaloids. **Table 1** summarizes some alkaloids that have been successfully synthesized in *in vitro* culture.

Alkaloids	Plant source	*In vitro* culture	Yield (%)	References
Indole alkaloids				
Serpentine	*Rauvolfia serpentina*	Cells suspension	0.03-0.06	[13]
		Plantlets	4.02-5.76	[15]
	Catharanthus roseus	Regenerated roots	0.06	[16]
Vinblastine	*Catharanthus roseus*	Regenerated roots	0.0005	[16]
Vincristine		Regenerated roots	0.01	[16]
Ajmalicine	*Catharanthus roseus*	Compact callus	0.003-0.005	[32]
		Cells suspension	0.16-0.7	[31, 36, 40, 57]
Reserpine	*Rauvolfia tetraphylla*	Callus	0.09-0.21	[14]
	Rauvolfia serpentina	Callus	0.1	[53]
Quinoline alkaloids				
Quinine	*Cinchona ledgeriana*	Hairy roots	0.058-0.23	[7, 8]
		Cells suspension	0.12-10.90	[12, 26]
	Cinchona officinalis	Hairy roots	0.05	[10]
Isoquinoline alkaloids				
Thebaine/codeine	*Papaver bracteatum*	Callus	0.01/0.0033	[17]
	Papaver somniferum	Embryogenic callus	0.0025/0.012	[84]
Berberine	*Coscinium fenestratum*	Cells suspension	0.004-0.02	[85]
	Thalictrum rugosum	Cells suspension	0.0015	[86]
Tropane alkaloids				
Atropine	*Datura metel*	Hairy roots	0.13-0.26	[60]
	Anisodus acutangulus	Hairy roots	0.0004-0.0007	[44]
Scopolamine/hyosciamine	*Brugmansia candida*	Hairy roots	0.24/0.06	[45]
Hyosciamine	*Hyosciamus muticus*	Callus	3.72-4.25	[61]
Terpenoid alkaloids				
Aconitine	*Aconitum napellus*	Cells suspension	0.035-0.043	[62]
Amaryllidaceae alkaloids				
Galanthamine	*Leucojum aestivum*	Plantlets	0.068–0.05	[41, 42, 58]
		Shoots	2.0–9.7	[37, 43]

Table 1. Report on some alkaloids produced in *in vitro* culture.

2.1. Biosynthesis pathway of alkaloids: different facts in cell cultures

TIAs are derived from amino acid tryptophan. **Figure 1** shows the branching pathways of many TIAs biosynthesis. Tryptophan combines with secologanin to form strictosidin. It seems that strictosidin is the intermediate base for many TIAs. From this stage, reserpine is synthesized. Strictosidin which is de-glycosylated to strictosidine aglycoside will convert to dehydrogeissoschizine. This compound in separate way via corynantheal will become cinchonidinone that by a reductive reaction forms cinchonidine and then quinine. Two

NADP-dependent enzymes specific to cinchona quinoline biosynthesis have been identified [20]. On the other hand, Alahmad *et al.* [21] reported that quinine is detected in some plants experiencing biotic stresses where quinine is presumably derived from phenolic compounds through the activity of polyphenol oxidase and/or phenol oxidase.

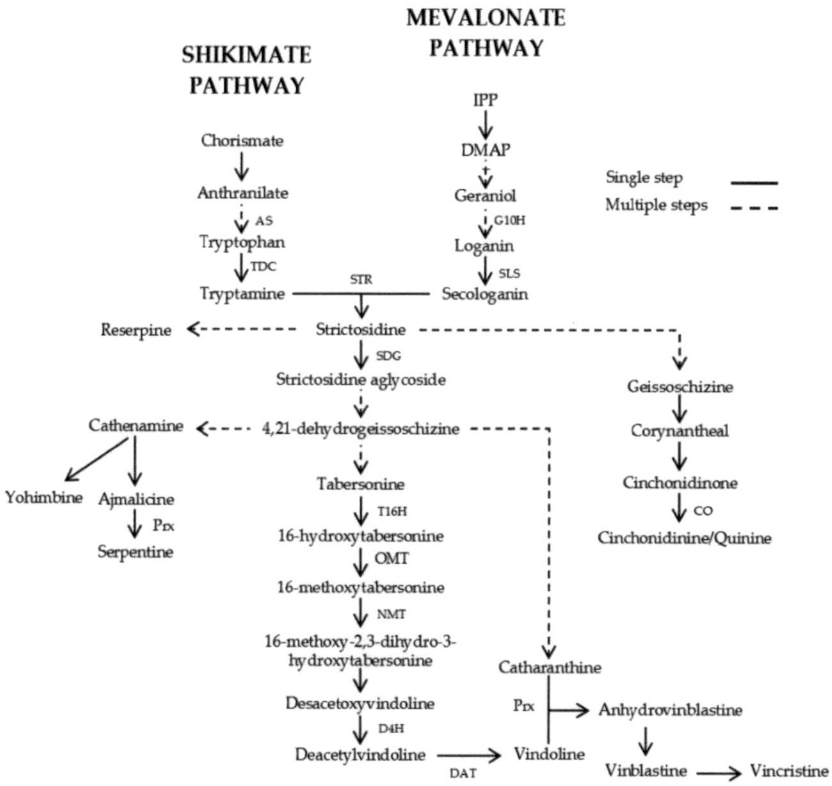

Figure 1. Biosynthesis pathways of terpenoid indole alkaloids (TIAs).

Dehydrogeissoschizine through six further steps via tabersonine pathway would arrive to form vindoline, with which catharanthine will be converted to bisindole alkaloids vinblastine and then vincristine. Dehydrogeissoschizine via cathenamine pathway produces yohimbine and ajmalicine. By oxidation, the latter compound becomes serpentine [20, 22, 23]. **Figure 2** demonstrates the basic molecular structures of strictosidine and its essential alkaloid derivatives.

The initial source for berberine biosynthesis is L-phenylalanine. From phenylalanine, through reticuline, coulerine, and canadine consecutively, berberine is produced. It belongs to the group benzyl isoquinoline alkaloids. The biosynthesis of atropine starting from L-phenylalanine too first undergoes a transamination forming phenylpyruvic acid which is then reduced to phenyllactic acid. The latter compound combines with co-A to become tropine. Following many further steps, it converts to hyoscyamine which is then racemized to atropine [24].

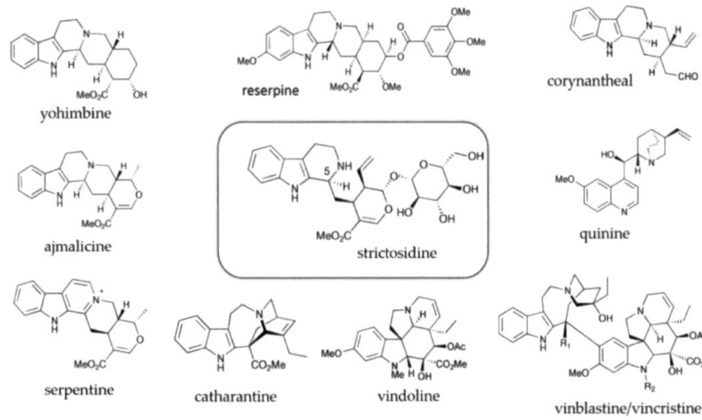

Figure 2. Strictosidine as central basic molecule of TIAs and its derivative molecules.

The biosynthesis of galanthamine also starts with enzymatic conversion of L-phenylalanine. The enzyme phenylalanine ammonialyase (PAL) converts it to protocatechuic aldehyde, and L-tyrosine, which includes the enzyme tyrosine decarboxylase producing tyramine. The junction of the tyramine and the protocatechuic aldehyde results in Schiff's base, which is converted to norbelladine. Norbelladine or related compounds undergo an oxidative coupling in Amaryllidaceae plants. 4-*O*-methylnorbelladine is considered as a key intermediate. The biosynthesis of galanthamine involves the phenol oxidative coupling of *O*-methylnorbelladine, and then following further steps, galanthamine is produced [25].

By identifying the pathway, the application of a precursor in culture media is a common practice. In *C. pubescens* cells, tryptophan as a precursor, fed in the culture media, did not improve quinoline alkaloids level [11, 12], but *C. ledgeriana* cells successfully increased the synthesis of cinchonidine by tryptophan feeding [26]. Culture media, with or without Cinchona cells in it, treated with tryptophan accumulated β-carboline alkaloids and indole-3-aldehyde [11, 27]. It indicates that tryptophan does not go along the pathway to quinoline alkaloids biosynthesis in Cinchona-cultured cells. Moreover, it was reported by [28] that simple β-carboline alkaloids such as tetrahydro-β-carboline-3-carboxylic acid and 1-methyl-tetrahydro-β-carboline-3-carboxylic acid are readily formed from tryptophan as well as from tryptamine by Pictet-Spengler reaction in food and beverages.

Feeding geraniol, 10-hydroxygeraniol, or loganin to a *C. roseus* hairy root culture resulted in significant increases in the accumulation of tabersonine [29], but the addition of tryptophan or tryptamine separately had no effect. It is presumed that only the early parts of the pathway are present in the hairy roots cultures where certain enzymes involved considerably in alkaloid production. An elicitor, jasmonic acid, combined with feeding either loganin or tryptamine did not further enhance the accumulation of indole alkaloids [30]. Precursors loganin, tryptamine, or their combination were fed to noninduced and methyl jasmonic acid (MeJA)-induced cultures. TIA production was not significantly enhanced in either noninduced or MeJA-induced cultures with precursor feeding. It seems that in noninduced cells, steps downstream of loganin and tryptamine were disturbed because of the accumulation of loganin or tryptamine in the cells with precursor feeding.

An increase in reserpine content was observed in 50 mg L^{-1} tryptophan fed callus culture than in other concentrations in *R. tetraphylla* callus [14]. Addition of secologanin increased the alkaloid ajmaline production 10-folds in *C. roseus* [31]. More results were reported [32] that succinic acid, tryptamine, and tryptophan feedings also significantly increased ajmalicine and catharanthine production by compact callus cultures of *C. roseus*, while geraniol feeding inhibited biomass and alkaloid accumulation. Treatment on the whole plant culture of *R. serpentina* with tryptamine (1.0 mg L^{-1}) alone resulted in enhanced reserpine production while the highest reserpine yield was obtained after 36 days of elicitation in cultures treated with 0.1 mmol salicylic acid and enriched with 1.0 mg L^{-1} tryptamine [15]. Research on the utilization of precursor to upscale *in vitro* production of other alkaloids is still limited, probably due to limited revelation of the biosynthesis mechanism.

2.2. Factors affecting alkaloids *in vitro* production

2.2.1. *Propagules as the initial plant material*

Theoretically all plant parts are capable of producing callus which then grows undifferentiated or differentiated. We have discussed above that some alkaloids are synthesized in particular organelles or organs, meaning that cytodifferentiation is required. An identification process for the most productive cell clones or propagules is important to be done, as the initial cultured plant material.

Cell type-specific localization applies also for berberine biosynthesis and accumulation which are temporally and spatially separated. In a berberine-producing plant, *Thalictrum flavum*, it was reported that the transcripts of its nine biosynthetic genes were confined to the pericycle and adjacent cortical cells of roots, and protoderm of leaf primordia, whereas berberine accumulation occurred in the root's endodermal cells, also in pith and cortex of rhizomes [33]. Based on this report, the use of root tissues or transformed hairy roots can be considered.

2.2.2. *Plant growth regulators*

Increasing or lowering the growth regulators in culture media gives impacts in alkaloids production. Growth regulators determine secondary metabolites biosynthesis as well as biomass production. It is well known that growth promoters such as auxins and cytokinins have important roles in plant growth and differentiation. However, it is also recognized that secondary metabolites are synthesized when the plant growth rate is decreasing, meaning when the plant starts to senesce or becomes stressed due to biotic or abiotic environment, secondary metabolites, including alkaloids, will appear. Based on this contradiction, a balanced treatment between growth promoters and stress-creating substances (growth inhibitors) has to be taken into account.

It was reported that auxins negatively influence alkaloid biosynthesis and accumulation at all levels in *C. roseus* cell culture [34]. Subculturing cells on an auxin-free medium results in increased *Tdc* and *Str* mRNA levels, while the addition of auxins rapidly decreases the *Tdc* mRNA. During the growth phase, 2,4-D strongly inhibits alkaloid production, but it recovers during the stationary phase. This is the reason that auxins are commonly added to the medium for callus induction, but they are added at a low concentration or omitted for the production of secondary metabolites. By contrast, the addition of cytokinin zeatin to an auxin-free *C. roseus* cell cultures resulted in an increase in ajmalicine alkaloid accumulation [35]. It was reported [36] that cytokinins and ethylene increased alkaloid accumulation in periwinkle callus or cell suspension cultures. It was mentioned that either exogenously applied cytokinins or ethylene (supplied by ethephon) greatly enhanced ajmalicine and serpentine accumulation in cells subcultured in a 2,4-D-free medium. Ethylene precursor, 1-aminocyclopropane-1-carboxylic acid, amplified galanthamine and lycorine content to sixfold in *Leucojum aestivum* while ethylene was reducing [37].

Abscisic acid (ABA) regulates various aspects of plant growth and development including seed maturation and dormancy, as well as adaptation to abiotic environmental stresses [38]. A report [39] stated that ABA stimulated accumulation of catharanthine and vindoline in *C. roseus*. Treatment of precursors fed *C. roseus* cells with ABA did not induce the accumulation of alkaloids but it delayed the catabolism of strictosidine [40]. In *C. ledgeriana* cell culture, ABA (1–3 mg L^{-1}) exhibited a growth promotion compared to the control medium containing picloram as an auxin source and BA, while the quinine content is comparable to that of ABA-free medium [12, 26], whereas paclobutrazol (PBZ) reduced the cell growth but induced significantly the quinine production [12].

There were variations in the accumulation of galanthamine in *L. aestivum in vitro* cultures grown in media with various combinations of growth regulators. In the absence of growth regulators, the amount of galanthamine was 0.0011% (DW). When cultivated in media with 10 μM α-naphthyl acetate and 0.5 μM BA, the highest yield was obtained, 0.0068% (DW) [41]. Cytokinin thidiazuron demonstrated better effect in *L. aestivum* shoot culture in which galanthamine biosynthesis achieved maximum yield of 0.05% (DW) [42].

2.2.3. Culture conditions

2.2.3.1. Nutrient and pH of culture media

Mineral and organic substances that are commonly incorporated in culture media are subject to modification. For galanthamine, the optimization of nitrate, ammonium, phosphate ions, and sucrose concentration increased the production in *L. aestivum* shoot culture [43]. The nitrogen concentration and NH_4^+/NO_3^- molar ratio of the culture medium often influence the synthesis of alkaloid. In *Anisodus acutangulus* hairy roots culture, cell growth and alkaloid yield were inhibited at low or high total nitrogen concentrations and the most favorable concentration of nitrogen for the maximum biomass and the highest tropane alkaloid yield was found to be 90 mM (NH_4^+/NO_3^- = 4:1). Regarding carbon source, medium containing sucrose, glucose, fructose, and galactose was compared; 3% sucrose gave the best yield in terms of cell growth and metabolism resulting in alkaloid productivity among different carbon sources used [44].

Medium component is not only the ingredients, but also pH. At pH 5.5 and at certain concentrations of acetic acid, the content of scopolamine and hyoscyamine increased in root cultures of *Brugmansia candida* and also promoted the release of both alkaloids into the medium. Lowering the pH to 3.5 and 4.5 reduced the accumulation of both alkaloids, but at a pH of 4.5, their release increased significantly. Acetic and citric acid stimulated the release of scopolamine and hyoscyamine, presumably due to permeability change of cell membrane [45]. A hairy root line, which was induced from leaves of *A. acutangulus*, was subjected to pH variation in the culture medium [44]. The biomass yield of hairy roots grown at pH 6.5 was two times higher than that at pH 4.5. However, the maximum production yield of tropane alkaloid was reached at pH 4.5.

2.2.3.2. *Light*

Light is an environmental factor most frequently reported having significant effect in secondary metabolite biosynthesis. Light can affect plant differentiation, morphology, and metabolic activities. Cultures that need certain stage of shoot differentiation to produce a secondary metabolite require light, indicating that some enzymes for the biosynthesis will be activated by light exposure. It was reconfirmed [46] that in lupine (*Lupinus* spp.), enzymes of quinolizidine alkaloids (QAs) and the precursor lysine are localized in the chloroplast. Consequently, the alkaloid formation requires cell differentiation to greening tissue and is governed by light. The conversion of ajmalicine to serpentine involves the role of peroxidase. Light-grown cell cultures of *C. roseus* had a 20-fold higher vacuolar peroxidase activity compared to those of dark-grown cells and the accumulation of serpentine in light-grown cells was higher [47]. Vindoline biosynthetic pathway is also regulated by light as proven with callus culture of *C. roseus* [32, 48]. Maximum galanthamine production was achieved in cultures grown under light conditions. Galanthamine in the cultures under light is more than twice (73.8 $\mu g\ g^{-1}$ DW) than those cultivated under dark conditions (38.5 $\mu g\ g^{-1}$ DW) [25].

It seems that alkaloids biosynthesis in Cinchona, at least partially, is light regulated. In contrast to the abovementioned cases, transformed cells of *C. ledgeriana* grown in the dark produced a remarkable enhancement of alkaloid accumulation, 50 times greater than those cultured under the light [49]. Blue light was found detrimental to alkaloid accumulation, but red or green light has the same effect with the darkness. Furthermore, alternating the dark and light periods for every 28 days resulted in alternate high and low alkaloid productivity, suggesting that this was not simply an adaptation effect.

2.2.3.3. *Temperature*

Cultivation of *C. roseus* hairy roots at different temperatures was found to have an effect on growth rate and indole alkaloid content. When lowering the temperature, the roots responded by increasing the degree of unsaturation of cellular lipids, which was mainly due to an increased proportion of linolenic acid. The modifications in lipid composition might be necessary for the roots to retain the proper cell membrane fluidity at each temperature. Although changes in membrane lipids might happen, the distribution of indole alkaloids between the roots and the medium was undetectable. Instead, the level of alkaloid accumulation in the roots increased significantly with lowering temperature [50].

2.3. Improvements to enhance *in vitro* alkaloids production

Plant alkaloids are usually produced in very low level, both in intact plant and in cell cultures. It depends greatly on the physiological and developmental stages of the plant or the plant cells. There are an increasing number of reports that plants and endophytic fungi produce secondary metabolites through mutualistic symbiosis. This issue is attracting, but the aseptic method of *in vitro* culture has to consider some modifications to benefice the involvement of endophytic microbes. Fungal endophytic *Curvularia* spp. and *Choanephora infundibulifera* have been found in *C roseus* enhancing vindoline content by 229–403% [51]. An example of producing a non-alkaloid substance, paclitaxel, has been reported [52]. Suspension cells of *Taxus chinensis* var. *mairei* was co-cultured with its endophytic fungi, *Fusarium mairei*, in a 20-L co-bioreactor for paclitaxel production. By using co-bioreactor that consists of two-unit tanks (10 L each) separated by a membrane, and then culturing *Taxus* suspension cells in one tank and the fungi in another, a desirable yield of paclitaxel was obtained in *Taxus* cell cultures. The co-cultured *Taxus* cell cultures produced 25.63 mg L^{-1} of paclitaxel within 15 days or equivalent to 1.71 mg L^{-1} per day and 38-fold higher than that by uncoupled culture (0.68 mg L^{-1} within 15 days).

Many other metabolic manipulations can be applied along with genetic engineering in attempting a higher production of desirable alkaloid compounds. To achieve an industrial scale of production, one has to obtain a stable, high-producing cell line of the plant of interest. At least two approaches of metabolic manipulation are being considered: (a) metabolic improvements such as screening and selection for high-producing cell lines and the stimulation of biosynthetic activities through various methods, and (b) optimization of growth and production medium.

2.3.1. Metabolic improvements

2.3.1.1. Cells screening

Screening of germplasm and selection methods to obtain highly productive cell clones is suggested. *R. serpentina* callus culture was divided into two cell strains of different fluorescence under 365-nm UV light [53]. The results indicated that the yellow-green fluorescent cell strain produced much more reserpine than the blue-white strain. Screening on the slow-growing cells of 13-year-old cultures of *R. serpentina* through four successive subcultures into a liquid medium has revived the cells. After three further generations, the cells recovered their growth rate and enhanced reserpine production [13].

Transformed hairy root cultures of *C. roseus* established approximately 150 transformants from four different cultivars. They were screened for desirable traits in growth and indole alkaloid production. Five hairy root clones grew well in liquid culture. The levels of alkaloids ajmalicine, serpentine, and catharanthine in these five clones were higher compared with cell suspensions reported elsewhere; the experiment also indicated the presence of vindoline in two clones at levels over three orders of magnitude, greater than the minute amounts reported in cell culture [54]. It was observed [55] from 11 cell lines of *C. roseus* derived from protoplast, 2 cell shapes, spherical and cylindrical. The production of ajmalicine and catharanthine was significantly greater when the cell aspect ratio (cell length/width) was more than 2.8.

2.3.1.2. Elicitors

In nature, a wide range of environmental stresses are threatening the plants. Secondary metabolites are frequently increased when the plant encounters to environmental stresses, biotic or abiotic. The phenomena found led us to believe that specific secondary metabolite functions as protective agents against the stresses. Based on those reasons, the use of artificial stresses or elicitors is common in *in vitro* cultures to enhance a desirable compound production. The most frequently used elicitors are high or low temperature, drought, medium salinity, growth retardants, microbial toxin, fungal carbohydrates, yeast extract, MeJA, and chitosan.

Polyethylene glycol (PEG) as a drought-creating agent incorporated in the callus culture media of *C. roseus* did not affect the production of vinblastine nor vincristine [56]. In contrast with ajmalicine, *C. roseus* cell suspension cultures treated with cadmium [57] resulted in the enhancement of ajmalicine content as well as TDC enzyme activity. Treatment with 250 mM mannitol and 4 g L^{-1} KCl as growth-stressing agents in compact callus culture of *C. roseus* yielded ajmalicine about four-fold higher than the control [32]. For quinine, paclobutrazol combined with mannitol in the media of *C. ledgeriana* suspension cultures significantly improved quinine content [12]. Galanthamine and lycorine were 1.36- and 1.67-fold higher compared to the control, respectively, achieved after elicitation with jasmonic acid in *L aestivum* cell culture [58]. The addition of MeJA at the start of the cultivation resulted in a two-fold increase in the concentration of galanthamine. Among the given elicitors, copper sulfate, silver nitrate, salicylic acid, and MeJA in *L. aestivum* shoot cultures [59], it was found that MeJA increased galanthamine biosynthesis the most.

Biotic (*Bacillus cereus* and *Staphylococcus aureus*) and abiotic ($AgNO_3$ and nanosilver) elicitors were added to the hairy root cultures of *Datura metel* [60]. All the elicitors influenced biomass accumulation and atropine production. Among the tested elicitors, nanosilver was the most effective in enhancing the hairy roots' atropine content. Chitosan increased the content of scopolamine and hyoscyamine in transformed roots of *B. candida* [45].

2.3.1.3. Optimization of culture conditions

Generally after having basic conditions required by cell cultures in regard to obtaining an acceptable cell growth rate containing certain level of alkaloids, it is necessary to optimize the culture media as well as the culture conditions to reach the maximum and consistent production of the desired substance. Some examples are given below.

Full-strength MS medium was the best for nourishing the growth of *Hyoscyamus muticus* cell cultures. The equal combination ratios of BA and NAA at 1.0 mg L^{-1} gave the highest biomass accumulation and alkaloid production of hyoscyamine. Culture grown under light was higher in their growth rate than that grown in the dark, but the alkaloid content was relatively higher under dark condition. Furthermore, low inoculum density of cell suspension cultures (25 g L^{-1} (FW) presented a shortest doubling time (2.93 days), while those obtained with a high inoculum (100 g L^{-1} (FW) reached 13.47 days. At inoculum density of 50 g L^{-1} (FW), the highest dry cell weight and alkaloid content were attained [61].

To increase aconitine alkaloid production in cell culture of *Aconitum napellus,* liquid MS medium supplemented with 1 mg L^{-1} and kinetin 0.1 mg L^{-1} gave the best growth rate and aconitine yield. Cell density of 3 g (FW) per flask resulted in the highest amount of cell biomass, and the addition of salicylic acid and yeast extract combined with 5% sucrose improved the alkaloid level 2.5–3-fold higher compared to that from the control media [62].

By modification of a standard MS culture medium and further optimization with the inclusion of lactose as the carbohydrate source, NAA and kinetin as growth regulators, a system has been developed for *C. roseus* cell suspension cultures to elevate both the growth and catharanthine alkaloid accumulation in a single-stage culture of 14–21 days [39].

Basic practices related to metabolic production process of any substance is pre-requisite for scaling-up by using of bioreactor. This means that certain level of alkaloids productivity has been achieved in general with particular method of cells/tissue cultures. A laboratory scale of bioreactor or biofermenter is a large vessel, 10 L of maximum capacity. Culturing cells suspension in bioreactor for a substance production needs a balanced combination of technical design and internal culture conditions that may be created. Control of medium pH, dissolved oxygen, flow rate of gas exchange, nutrient supply, cells density, agitation speed, as well as temperature are among the factors of custodian importance [63, 64].

2.3.2. Genetic engineering

Many achievements have been made in modifying various metabolic pathways by using specific genes encoding biosynthetic enzymes or regulatory proteins. Full-length cDNAs namely *CrRR2* and *CrRR3* originated from a *C. roseus* cDNA library have been introduced into cell suspension culture *of C roseus. CrRR2* gene was expressed at a very low level, if any, that treatment by cytokinins did not trigger its transcription. The *CrRR3* gene expression is root-specific and the transcripts are transiently up-regulated. In cells suspension, transcript amounts of *CrRR3* remained unchanged even by the treatment with NaCl, abscisic acid, or jasmonic acid, but was increased with 10^{-6} M trans-zeatin [65].

In *C. officinalis* hairy roots [10], a binary plant vector construct whose T-DNA region contained constitutive-expression versions (CaMV35S promoter with double enhancer and *nos* terminator) of *tdc* and *str* cDNA clones was used from *C. roseus*. It was then combined with an intron possessing the β-glucuronidase (*gus*-int) reporter gene and a hygromycin phospho-transferase (*hpt*) selection gene. The products of TDC and STR, tryptamine and strictosidine, were found in high amounts, 1200 and 1950 mg g^{-1} (DW), respectively. Quinine and quinidine levels were found to rise up to 500 and 1000 mg g^{-1} (DW), respectively. However, the genes were not expressed anymore one year after. They had completely lost their capacity to accumulate alkaloids.

http://dx.doi.org/10.5772/66288

Based on the knowledge of biosynthetic pathways, a logic strategy can be made. Some efforts to induce up-regulation of the transcript levels of key pathway genes are also commonly performed, but the successes are not universal. In cultures of *C. roseus* elicited with MeJA (250 μM) on day 21, *g10h, tdc, str*, and *sgd* expression increased by three- to nine-fold after 24 h. Up-regulation of the gene expression was followed by a 160, 440, and 420% increase in strictosidine, ajmalicine, and tabersonine levels, respectively, after 5 days [30].

Genetic engineering is still promising for elaboration since the real problems encountered in stabilizing the alkaloids production after transformation or induction rest to be elucidated.

3. Alkaloids accumulation in cultured cells

The fact is that alkaloids are synthesized and accumulated in plants to cope with unfavorable environment. They hold adaptive significance to survive in diverse ecological situation. But a quite number of these compounds are toxic and can be dangerous to the producing plant itself [66]. Consequently, to maintain homeostasis, feedback inhibition of certain enzymes in the biosynthesis pathway can occur in particular cell compartment that leads to cease further alkaloids production. As a prerequisite for the defense function, however, the sequestration of a respective alkaloid to a critical amount is necessary, then the compound is frequently confined in vacuoles of specialized cells, as a means of detoxification.

3.1. Secretory structures

Plants have various specialized cells where a wide variety of products are often synthesized and/or stored, namely secretory cells. External secretory structures include trichome, gland, and nectary, but alkaloids are found mostly in internal structures such as idioblast, laticifer, or secretory cavity and canal [67]. They are isolated cells that differ from the surrounding cells.

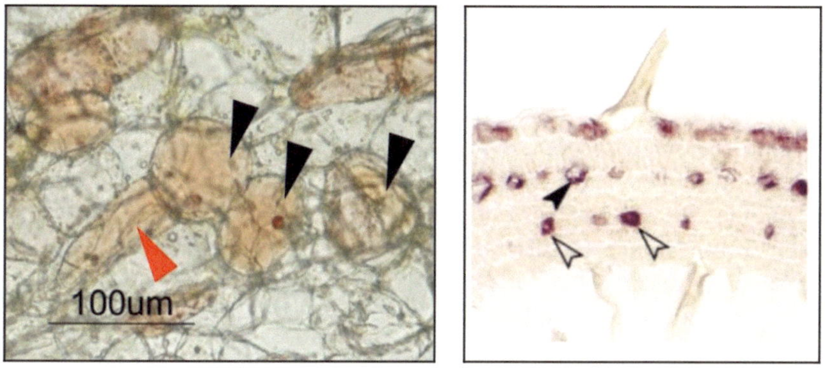

Figure 3. Spheric and elongated laticifers containing alkaloids among the cell aggregates cultures (left); idioblasts (open arrowheads) and laticifer (solid arrowhead) in an intact leaf (right) of *Catharanthus roseus* (left: courtesy of Iskandar and Iriawati [56]; right: with permission of the American Society of Plant Physiologists from [48]).

It was reported [48, 68] that vinblastine and vincristine are confined in idioblast and laticifer cells located among the palisade and spons cells, whereas catharanthine is found in leaf cuticle, and specialized cells of cortex [69]. According to [70], the capacity of alkaloid synthesis in C. *roseus* depends on the number of cells accumulating the alkaloid compounds. The genes for the enzymes involved in the final stages of those two bisindole alkaloids synthesis, D4H and DAT, are only expressed and activated within the idioblast and laticifer. Those cells are also the site of the secondary metabolite accumulation [48, 71, 72]. **Figure 3** demonstrates the existence of secretory cells of C. *roseus* in callus culture and in intact leaf, suggesting that in *in vitro* cultures, such a structure can be regenerated.

In *Papaver somniferum*, three different sites have been identified; benzylisoquinoline alkaloids were stored in laticifers, the biosynthetic gene transcripts were restricted to companion cells, and the enzymes were detected in sieve elements [73].

Cinchona bark does not exhibit any specialized structure that may be the site of biosynthesis and/or storage of quinoline alkaloids. It was suggested [74] that quinoline alkaloid is located most probably in intercellular spaces of the bark, which are schizogenous, lysigenous, or schizo-lyzigeneous secretory ducts or canals. These are tube-like structures containing and producing secretions. No report on secretory structures in cultured Cinchona cells is found so far.

3.2. Alkaloids accumulation and harvest

Three factors affecting the yield of secondary metabolite should be considered, for example, biosynthesis of the product, accumulation of the product, and breakdown of the product. The actual yield will be the result of biosynthesis minus breakdown of the product, which provides storage in a certain cellular compartment as a mechanism of avoiding autointoxication of the producing cells [75].

Isoquinoline alkaloids, for example, berberine, are also synthesized in small vesicles which later fuse with tonoplast to release the alkaloids content into the vacuole. Both *Coptis japonica* and *Thalictrum flavum* cells, which have an ability to synthesize berberine, took up exogenous berberine from the culture medium to accumulate it exclusively in vacuoles. By contrast, *T. minus* cells, which excrete indigenous berberine mostly into the medium, did not take up exogenously supplied berberine [76]. Vacuole seems to serve as particular compartment to store high concentration of secondary metabolites, since some of them might be toxic for the producing plant itself when they exist in higher amount than needed for particular function, for instance in the defense or signaling [6]. A study on *C. ledgeriana* reconfirmed this [77]. The toxicity of cinchona alkaloids to cell cultures was related to alkaloid uptake from the medium with the purpose for selecting high-yielding cell lines. Quinine, the strongest, was completely toxic at 5.5 mM. Both quinine and quinidine were more toxic than their unmethoxylated precursors, cinchonidine and cinchonine.

Elimination of feedback inhibition of metabolic enzymes and inhibition of membrane transport can be made by certain products newly synthesized in a second phase introduced into liquid culture medium. Organ culture often demonstrates separate sites of synthesis and storage of secondary metabolites [78] which facilitates to avoid a saturated condition of a substance. Some *in vitro*–cultured cells have also different subcellular compartments for biosynthesis and storage of secondary metabolites. Increasing the permeabilization of vacuolar and cell membranes would let the metabolites excretion to occur into the medium or into the apoplast. Some chemicals have been used to alter membrane permeabilization,. Cd ranging from 0.05 to 0.4 mM for 24–48 h increased ajmalicine production and excretion into culture medium, particularly for cells at the mid-exponential growth phase [57], Tween 20 added for 24 h to the medium, after 2–4 weeks, the hairy roots line of D. *metel* culture excreted scopolamine into the culture medium, from 8.7 to 70% [79], or Tween 80 in roots culture of *D. inoxia* elevated hyoscyamine release in the medium [80].

3.3. Industrial-scale production of alkaloids by cell cultures

The high production cost to extract a secondary metabolite that creates the consequence of high price of the product has been described. The small taxol yield of *Taxus* species requires 10,000 kg of bark to produce 1 kg of taxol [81]. In 1993, it was reported that quinine and quinidine were produced at 300–500 metric tons yearly, which were extracted from 5000 to 10,000 metric tons of Cinchona bark [75]. Concern was raised about the environmental impact of the sourcing, leading to search for alternative sources of the alkaloid. In the end, an alternative environment-sustainable approach using plant cell culture has been developed.

Large-scale production of plant secondary metabolites through plant cells culture in bioreactors is technically feasible [64, 75]. However, the cost-benefit of such a production is still the major barrier. For some pricey products, it is still feasible. Another obstacle is that unfortunately some of the most interesting products exist only in very small amounts or are not at all produced in plant cell cultures. Hence, for industrial production, two crucial questions were raised [75], whether plant cells can be grown in large fermenters and whether the price of a product from such a large-scale plant cell culture will be competitive with conventional production methods. The questions are still relevant nowadays for some extent. The answers remain case-dependent, as during these 30 years several industrial-scale productions of secondary substances indeed have been developed [64].

In another group of secondary metabolite paclitaxel, Phyton Biotech [81] and Samyang Genex Corporation [82], both are the largest paclitaxel producer, have successfully increased the productivity of taxanes by using plant tissue culture in large scale fermenters and through the application of several strategies. By selecting high-yielding cell lines, Mitsui group produced berberine on a large scale with a productivity of 1.4 g L^{-1} over 2 weeks [83].

4. Conclusion

The facts that plant alkaloids can be produced by cultured cells have proven that cell culture system is feasible to produce enhanced yield of desirable substances as an alternative for natural sources. Plant cells have plasticity in some extent and are subject to manipulations.

Therefore, substantial advanced and continuous researches still need to be established, particularly in the biosynthetic pathway, the enzymes involved and their properties, the sites of each reaction step, the factors affecting those steps of biosynthesis and accumulation, and the consistency, as well as factors affecting the respective genes' expression in cultured cells. Specific cell type of a plant species necessitates a detailed exploration for its requirements to produce certain alkaloids.

Collective reports from all over the world will be likely to improve the basic knowledge as well as the technical approaches that will promote the development of a stable cells system serving as technological matter. Industrial production of secondary metabolites in bioreactors is not an impossible matter in the near future. It has been developing with steady progress, particularly for certain alkaloids.

Author details

Diah Ratnadewi

Address all correspondence to: dratnadewi@ipb.ac.id

Department of Biology, Faculty of Mathematics and Natural Sciences, Bogor Agricultural University, Bogor, Indonesia

References

[1] Kurian A, Sankar MA. Medicinal plants. In: Peter KV, editor. Horticultural Science Series-2. New Delhi, India: New India Publication Agency; 2007.

[2] Andrea G. Reserpine: The Treatment of Hypertension as a Cause of Depression [Internet]. 2012. Available from: http://flipper.diff.org/app/items/3998 [Accessed: 2016-07-21].

[3] Deshmukh SR, Ashrit DS, Patil BA. Extraction and evaluation of indole alkaloids from *Rauvolfia serpentina* for their antimicrobial and antiproliferative activities. International Journal of Pharmacy and Pharmaceutical Sciences. 2012;4(5):329-334.

[4] Kumari R, Rathi B, Rani A, Bhatnagar S. *Rauvolfia serpentina* L. Benth. Ex Kurz. phytochemical, pharmacological and therapeutic aspects. International Journal of Pharmaceutical Sciences Review and Research. 2013;23(2):348-355.

[5] Almagro L, Fernandez-Perez F, Pedreno MA. Indole alkaloids from *Catharanthus roseus*: bioproduction and their effect on human health. Molecules. 2015;20:2973-3000. DOI: 10.3390/molecules20022973

[6] Wink M, Roberts MF. Comparison of alkaloid synthesis, transport, and storage. In: Roberts MF, Wink M, editors. Alkaloids: Biochemistry, Ecology, and Medicinal Applications. New York, NY: Plenum Press; 1998. pp. 239-262.

[7] Hamill JD, Robins RJ, Rhodes MJC. Alkaloid production by transformed root cultures of *Cinchona ledgeriana*. Planta Medica. 1989;55:354-357.

[8] Toruan-Mathius N, Haris N, Santoso J, Ade-Heri. Effect of elicitation on growth and alkaloid quinoline production in hairy root of cinchona plant (*Cinchona succirubra* Pavon ex Klotzsch). Menara Perkebunan. 2006;74:10-22.

[9] Pasquali G, Goddijn OJM, De Waal A, Verpoorte R, Schiperoort RA, Hoge JHC, et al. Coordinated regulation of two indole alkaloid biosynthetic genes from *Catharanthus roseus* by auxins and elicitors. Plant Molecular Biology. 1992;18:1121-1131.

[10] Geerlings A, Hallard D, Caballero AM, Cardoso IL, van der Heijden R, Verpoorte R. Alkaloid production by *Cinchona officinalis* 'Ledgeriana' hairy root culture containing constitutive expression construct of tryptophan decarboxylase and strictosidine synthase cDNA from *Catharanthus roseus*. Plant Cell Reports. 1999;19:191-196.

[11] Robins RJ, Hanley AB, Richards SR, Fenwick GR, Rhodes MJC. Uncharacteristic alkaloid synthesis by suspension cultures of *Cinchona pubescens* fed with L-tryptophan. Plant Cell, Tissue and Organ Cultures. 1987;9(1):49-59. DOI: 10.1007/BF00046079

[12] Ratnadewi D, Satriawan D, Sumaryono. Enhanced production level of quinine in cell suspension culture of *Cinchona ledgeriana* Moens by paclobutrazol. Biotropia. 2013;20(1):10-18. DOI: 10.11598/btb.2013.20.1.10

[13] Yamamoto O, Yamada Y. Production of reserpine and its optimization in cultured *Rauvolfia serpentina* Benth. cells. Plant Cell Reports. 1986;5:50-53.

[14] Anitha S, Kumari BDR. Stimulation of reserpine biosynthesis in the callus of *Rauwolfia tetraphylla* L. by precursor feeding . African Journal of Biotechnology. 2006;5(8):659-661.

[15] Panwar GS, Guru SK. Stimulation of reserpine production in the whole plant culture of *Rauvolfia serpentina* L. by elicitors and precursor feeding. Journal of Biochemistry and Biotechnology. 2015;24(1):49-55. DOI: 10.1007/s13562-013-0235-5

[16] Ataei-Azimi A, Hashemloian BD, Ebrahimzadeh H, Majd A. High *in vitro* production of anti-cancer indole alkaloids from periwinkle (*Catharanthus roseus*) tissue culture. African Journal of Biotechnology. 2008;7(16):2834-2839. DOI: 10.5897/AJB08.110

[17] Kamo KK, Mahlberg PG. Morphinan alkaloids: Biosynthesis in plant (*Papaver* spp.) tissue cultures. In: Bajaj YPS, editor. Biotechnology in Agriculture and Forestry: Medicinal and Aromatic Plants I. 4th ed. Berlin Heidelberg New York, NY: Springer; 1988. p. 251-263.

[18] Yoshikawa T, Furuya T. Studies on plant tissue culture: Morphinan alkaloid production by tissues, differentiated from cultured cells of *Papaver somniferum*. Planta Medica. 1985;2:110-113. DOI: 10.1055/s-2007-969421.

[19] Smetanska I. Production of secondary metabolites using plant cell cultures. Advances in Biochemical Engineering/Biotechnology. 2008;111:187-228. DOI: 10.1007/10_2008_103.

[20] Isaac JE, Robins RJ, Rhodes MJC. Cinchoninone: NADP oxidoreductases I and II-novel enzymes in the biosynthesis of quinoline alkaloids in *Cinchona ledgeriana*. Phytochemistry. 1987;26(2):393-399.

[21] Alahmad K, Gasmalla MAA, Alyousef H. Efficient methods for polyphenol oxidase production. International Journal of Nutrition and Food Sciences. 2015;4(6):656-659. DOI: 10.11648/j.ijnfs.20150406.19

[22] O'Connor SE, Maresh JJ. Chemistry and biology of monoterpene indole alkaloid biosynthesis. Natural Product Reports. 2006;23:532-547. DOI: 10.1039/b512615k

[23] Zhu J, Wang M, Wen W, Yu R. Biosynthesis and regulation of terpenoid indole alkaloids in *Catharanthus roseus*. Pharmacognosy Reviews. 2015;9(17):24-28. DOI: 10.4103/0973-7847.156323

[24] Dewick PM. Medicinal Natural Products: A Biosynthetic Approach. 2nd ed. Chichester: John Wiley & Sons; 2002. 507 p.

[25] Berkov S, Ivanov I, Georgiev V, Codina C, Pavlov A. Galanthamine biosynthesis in plant *in vitro* systems. Engineering in Life Sciences. 2014;(14):643-650. DOI: 10.1002/elsc.201300159

[26] Ratnadewi D, Sumaryono O. Quinoline alkaloids in suspension cultures of *Cinchona ledgeriana* treated with various substances. Hayati Journal of Biosciences. 2010;17(4):179-182. DOI: 10.4308/hjb.17.4.179

[27] Harkes PAA, De Jong PJ, Wijnsma R, Verpoorte R, Van der Leer. The influence of production media on cinchona cell cultures: spontaneous formation of β-carbolines from L-tryptophan. Plant Sciences. 1986;47:71-76. DOI: 10.1016/0168-9452(86)90012-9

[28] Cao R, Peng W, Wang Z, Xu A. B-carboline alkaloids: biochemical and pharmacological functions. Current Medicinal Chemistry. 2007;14(4):479-500.

[29] Morgan JA, Shanks JV. Determination of metabolic rate-limitations by precursor feeding in *Catharanthus roseus* hairy root cultures. Journal of Biotechnology. 2000;79:137-145.

[30] Glokany S, Loring RH, Glick J, Lee-Parsons CW. Assessing the limitations to terpenoid indole alkaloid biosynthesis in *Catharanthus roseus* hairy root cultures through gene expression profiling and precursor feeding. Biotechnology Program. 2009;25(5):1289-1296.

[31] Merillon JM, Doireau P, Guillot A, Chenieux JC, Rideau M. Indole alkaloid accumulation and tryptophan decarboxylase activity in *Catharanthus roseus* cells cultured in three different media. Plant Cell Reports. 1986;5:23-26.

[32] Zhao J, Hu Q, Guo Y-Q, Zhu W-H. Effects of stress factors, bioregulators, and synthetic precursors on indole alkaloid production in compact callus clusters cultures of *Catharanthus roseus*. Applied Microbiology and Biotechnology. 2001;55:693-698. DOI: 10.1007/s002530000568.

[33] Samanani N, Park SU, Facchini PJ. Cell type-specific localization of transcripts encoding nine consecutive enzymes involved in protoberberine alkaloid biosynthesis. Plant Cell. 2005;17(3):915-926. DOI: 10.1105/tpc.104.028654

[34] Arvy MP, Imbault N, Naudascher F, Doireau P. 2,4-D and alkaloid accumulation in periwinkle cell suspensions. Biochimie. 1994;76(5):410-416. DOI: 10.1016/0300-9084(94)90117-1

[35] Decendit A, Liu D, Ouelhazi L, Doireau P, Merillon JM, Rideau M. Cytokinin-enhanced accumulation of indole alkaloids in *Catharanthus roseus* cell cultures: the factors affecting the cytokinin response. Plant Cell Reports. 1992;11:400-403.

[36] Yahia A, Kevers C, Gaspar T, Chenieux JC, Rideau M, Creche J. Cytokinins and ethylene stimulate indole alkaloid accumulation in cell suspensions cultures of *Catharanthus roseus* by two distinct mechanisms. Plant Sciences. 1998;133:9-15.

[37] Ptak A, Tahchy AE, Wyzgolik G, Henry M, Laurain-Mattar D. Effects of ethylene on somatic embryogenesis and galanthamine content in *Leucojum aestivum* L. cultures. Plant Cell, Tissue and Organ Cultures. 2010;102(1):61-67. DOI: 10.1007/s11240-010-9706-8

[38] Beaudoin N, Serizet C, Gosti F, Giraudat J. Interactions between abscisic acid and ethylene signaling cascades. The Plant Cell. 2000;12(7):1103-1115. DOI: 10.1105/tpc.12.7.1103

[39] Smith JI, Smart NJ, Kurz WGW, Misawa M. The use of organic and inorganic compounds to increase the accumulation of indole alkaloids in *Catharanthus roseus* (L.) G. Don cell suspension cultures. Journal of Experimental Botany. 1987;38(9):1501-1506.

[40] El-Sayed M, Verpoorte R. Effect of phytohormones on growth and alkaloid accumulation by a *Catharanthus roseus* cell suspension cultures fed with alkaloid precursors tryptamine and loganin. Plant Cell, Tissue and Organ Culture. 2002;68(3):265-270. DOI: 10.1023/A:1013968826635

[41] Diop MF, Ptak A, Chretien F, Henry M, Chapleur J, Laurain-Mattar D. Galanthamine content of bulbs and *in vitro* cultures of *Leucojum aestivum*. Natural Product Communications. 2006;1:475-479.

[42] Ptak A, Simlat M, Kwiecie'n M, Laurain-Mattar D. *Leucojum aestivum* plants propagated in *in vitro* bioreactor culture and on solid media containing cytokinins. Engineering in Life Sciences. 2013;13:261-270. DOI: 10.1002/elsc.201200109.

[43] Georgiev V, Berkov S, Georgiev M, Burrus M, Codina C, Bastida J, *et al.* Optimized nutrient medium for galanthamine production in *Leucojum aestivum* L. *in vitro* shoot system. Zeitschriff fur Naturforschung. 2009;64(c):219-224.

[44] Liu Q, Cui L, Guo Y, Ni X, Zhang Y, Kai G. Optimization of nutritive factors in culture media for growth and tropane alkaloid production from *Anisodus acutangulus* hairy roots. Journal of Applied Pharmaceutical Science. 2013;3(01):001-004. DOI: 10.7324/JAPS.2013.30101

[45] Pitta-Alvarez SI, Giulietti AM. Influence of chitosan, acetic acid and citric acid on growth and tropane alkaloid production in transformed roots of *Brugmansia candida*: effect of medium pH and growth phase. Plant Cell, Tissue and Organ Culture. 1999;59:31-38. DOI: 10.1023/A:1006359429830.

[46] Wink M, Hartmann T. Enzymatic synthesis of quinolizidine alkaloid esters: a tigloyl-CoA:13-hydroxylupanine O-tigloyltransferase from *Lupinus albus*. Planta. 1982;156: 560-565.

[47] El-Sayed M, Verpoorte R. Catharanthus terpenoid indole alkaloids: biosynthesis and regulation. Phytochemistry Reviews. 2007;6(2):277-305. DOI: 10.1007/s11101-006-9047-8

[48] St Pierre B, Vazquez-Flota FA, De Luca V. Multicellular compartmentation of *Catharanthus roseus* alkaloid biosynthesis predicts intercellular translocation of a pathway intermediate. The Plant Cell. 1999;11:887-900.

[49] Payne J, Rhodes MJ, Robins RJ. Quinoline alkaloid production by transformed cultures of *Cinchona ledgeriana*. Planta Medica. 1987;53(4):367-372. DOI: 10.1055/s-2006-962740

[50] Toivonen L, Laakso S, Rosenqvist H. The effect of temperature on hairy root cultures of *Catharanthus roseus*: growth, indole alkaloid accumulation and membrane lipid composition. Plant Cell Reports. 1992;11(8):395-399.

[51] Pandey SS, Singh S, Babu CSV, Shanker K, Srivastava NK, Shukla AK, *et al.* Fungal endophytes of *Catharanthus roseus* enhance vindoline content by modulating structural and regulatory genes related to terpenoid indole alkaloid biosynthesis. Scientific Reports. 2016;6:1-14. DOI: 10.1038/srep26583

[52] Li YC, Tao WY, Cheng L. Paclitaxel production using co-culture of Taxus suspension cells and paclitaxel-producing endophytic fungi in a co-bioreactor. Applied Microbiology and Biotechnology. 2009;83:233-239. DOI: 10.1007/s00253-009-1856-4

[53] Yamamoto O, Yamada Y. Selection of reserpine producing cell strain using UV-light and optimization of reserpine production in the selected cell strain. Plant Cell, Tissue and Organ Culture. 1987;8:125-133.

[54] Bhadra R, Vani S, Shanks JV. Production of indole alkaloids by selected hairy root lines of *Catharanthus roseus*. Biotechnology and Bioengineering. 1993;41(5):503-602.

[55] Kim SW, Jung KH, Kwak SS, Liu JR. Relationship between cell morphology and indole alkaloid production in suspension cultures of *Catharanthus roseus*. Plant Cell Reports. 1994;14:23-26. DOI: 10.1007/BF00233292.

[56] Iskandar NN, Iriawati. Vinblastine and vincristine production on Madagascar periwinkle (*Catharanthus roseus* (L.) G. Don) callus culture treated with polyethylene glycol. Makara Journal of Science. 2016;20(1):7-16. DOI: 10.7454/mss.v20i1.5656

[57] Zheng Z, Wu M. Cadmium treatment enhances the production of alkaloid secondary metabolites in *Catharanthus roseus*. Plant Science. 2004;166(2):507-514.

[58] Ivanov I, Georgiev V, Pavlov A. Elicitation of galanthamine biosynthesis by *Leucojum aestivum* liquid shoot cultures. Journal of Plant Physiology. 2013;170:1122-1129.

[59] Schumann A, Berkov S, Claus D, Gerth A, Bastida J, Codina C. Production of galanthamine by *Leucojum aestivum* shoot grown in different bioreactor systems. Applied Biochemistry and Biotechnology. 2012;167(7):1907-1920. DOI: 10.1007/s12010-012-9743-3

[60] Shakeran Z, Keyhanfar M, Ashgari G, Ghanadian M. Improvement of atropine production by different biotic and abiotic elicitors in hairy root cultures of *Datura metel*. Turkish Journal of Biology. 2015;39:111-118. DOI: 10.3906/biy-1405-25

[61] Aly UI, El-Shabrawi HM, Hanafy M. Impact of culture conditions on alkaloid production from undifferentiated cell suspension cultures of Egyptian henbane. Australian Journal of Basic and Applied Sciences. 2010;4(10):4717-4725.

[62] Hwang SJ, Kim YH, Pyo BS. Optimization of aconitine production in suspension cell cultures of *Aconitum napellus* L. Korean Journal of Medicinal Crop Science. 2004;12(5): 366-371.

[63] Ruffoni B, Pistelli L, Bertoli A, Pistelli L. Plant cells cultures: bioreactors for industrial production. In: Giardi MT, Rea G, Berra B, editors. Bio-Farms for Nutraceuticals: Functional Food and Safety Control by Biosensors. Germany: Landes Bioscience and Springer+Business Media; 2010. p. 203-221.

[64] Georgiev MI, Weber J. Bioreactors for plant cells: optimization as tools for wider commercialization hardware configuration and internal environment. Biotechnology Letters. 2014;36(7):1359-1367. DOI: 10.1007/s10529-014-1498-1

[65] Papon N, Oudin A, Vansiri A, Rideau M, CheAnieux J-C, Creche J. Differential expression of two type-A response regulators in plants and cell cultures of *Catharanthus*

roseus (L.) G. Don. Journal of Experimental Botany. 2003;54(388):1793-1795. DOI: 10.1093/jxb/erg185

[66] Wink M. The plant vacuole: a multifunctional compartment. Journal of Experimental Botany. 1993;44:231-246.

[67] Dickison WC. Integrative Plant Anatomy. CA, USA: Academic Press; 2000. 533 p.

[68] Carqueijeiro I. Unraveling the metabolism and transmembrane transport of the highly valuable medicinal alkaloids from *Catharanthus roseus* [dissertation]. Portugal: Faculdade de Ciencia Universidade do Porto; 2012. 208 p. Available from: https://repositorio-aberto.up.pt/bitstream/10216/68178/2/11945.pdf

[69] Eilert U, Wolters B, Constabel F. Ultrastructure of acridone alkaloid idioblasts in roots and cell cultures of *Ruta graveolens*. Journal of Canadian Botany. 1985;64:1089-1096. DOI: 10.1139/b86-149

[70] Endress R. Plant Cell Biotechnology. Berlin, Heidelberg: Springer Verlag; 1994. 330 p. DOI: 10.1007/978-3-662-02996-1

[71] Rischer H, Oresic M, Seppanen-Laakso T, Katajamaa M, Lammertyn F, Ardiles-Diaz W, *et al.* Gene-to-metabolites networks for terpenoid indole alkaloid biosynthesis in *Catharanthus roseus* cells. PNAS. 2006;103(14):5614-5619. DOI: 10.1073/pnas.0601027103

[72] Zhou ML, Hou HL, Zhu XM, Shao JR, Wu YM, Tang YX. Molecular regulation of terpenoid indole alkaloids pathway in the medicinal plant, *Catharanthus roseus*. Journal of Medicinal Plant Research. 2010;4(25):2760-2772.

[73] Bird DA, Franceschi VR, Facchini PJ. A tale of three cell types: alkaloid biosynthesis is localized to sieve elements in opium poppy. Plant Cell. 2003;15(11):2626-2635. DOI: http://dx.doi.org/10.1105/tpc.015396

[74] Metwaly A. Bark, Wood and Galls. Bark 1-Cinchona [Internet]. 2015 . Available from: http://www.slideshare.net/AhmedMetwaly3/bark-1 [Accessed: 2016-08-17]

[75] Verpoorte R, Van der Heijden R, Schripsema J, Hoge JHC, Ten Hoopen HJG. Plant cell biotechnology for the production of alkaloids: present status and prospects. Journal of Natural products. 1993;56(2):186-207. DOI: 10.1021/np50092a003

[76] Sato H, Kobayashi Y, Fukui H, Tabata M. Specific differences in tolerance to exogenous berberine among plant cell cultures. Plant Cell Reports. 1990;9(3):133-136. DOI: 10.1007/BF00232088

[77] Walton NJ, Parr AJ, Robins RJ, Rhodes MJ. Toxicity of quinoline alkaloids to cultured *Cinchona ledgeriana* cells. Plant Cell Reports. 1987;6(2):118-121. DOI: 10.1007/BF00276667

[78] Dornenburg H, Knorr D. Strategies for the improvement of secondary metabolite production in plant cell cultures. Enzyme and Microbial Technology. 1995;17(8): 674-684. DOI: 10.1016/0141-0229(94)00108-4

[79] Cusido RM, Palazon J, Pinol MT, Bonfill M, Morales C. *Datura metel: in vitro* production of tropane alkaloids. Planta Medica. 1999;65(2):144-148. DOI: 10.1055/s-1999-13976

[80] Boitel-Conti M, Laberche JC, Lanoue A, Ducrocq C, Sangwan NB. Influence of feeding precursors on tropane alkaloid production during an abiotic stress in *Datura inoxia* transformed roots. Plant Cell, Tissue and Organ Cultures. 2000;60:131-137.

[81] Isah T. Natural sources of taxol. British Journal of Pharmaceutical Research. 2015;6(4): 214-227. DOI: 10.9734/BJPR/2015/16293

[82] Zhong JJ. Plant cell culture for production of paclitaxel and other taxanes. Journal of Bioscience and Bioengineering. 2002;94(6):591-599. DOI: 10.1016/S1389-1723(02) 80200-6

[83] Hussain MS, Fareed S, Ansari S, Rahman MS, Ahmad IZ, Saeed M. Current approaches toward production of secondary plant metabolites. Journal of Pharmacy & Bioallied Sciences. 2012;4(1):10-20. DOI: 10.4103/0975-7406.92725

[84] Oluk EA. Alkaloid production in tissue cultures of *Papaver somniferum* L. cv. Office-95. Plant Cell Culture & Biotechnology. 2006;16(1):1-4.

[85] Narasimhan S, Nair GM. Effect of auxins on berberine biosynthesis in cell suspension culture of *Coscinium fenestratum* (Gaertn.) Colebr-a critically endangered medicinal liana of Western Ghats. Indian Journal of Experimental Biology. 2004;42:616-619.

[86] Gugler K, Funk C, Brodelius P. Elicitor-induced tyrosine decarboxylase in berberine-synthetizing suspension cultures of *Thalictrum rugosum*. European Journal of Biochemistry. 1988;170:661-666.

Chapter 2

Synthesis, Structure and Biological Activity of Ephedra Heterocycles

Alejandro Cruz, Itzia I. Padilla-Martínez and
María Esther Bautista-Ramírez

Additional information is available at the end of the chapter

http://dx.doi.org/10.5772/67387

Abstract

Ephedra compounds are well known due to their biological activity. They have been widely used in asymmetric synthesis during the last decades. Recently, we have prepared reviews about the synthesis of acyclic and heterocyclic ephedra derivative compounds reported in the literature. In this chapter, the synthetic methodology to access acyclic and heterocyclic compounds derived from ephedra alkaloids and its structural analysis are discussed, included those due to the substitution of the hydroxy group by chlorine, sulfur, selenium, or nitrogen atoms. Biological activity analysis of some synthesized compounds was done, and some of them have displayed biological activity.

Keywords: ephedrines, chirality, configuration, biological activity, stereoespecificity

1. Introduction

Chirality in biological systems is of main significance since in enzymes and drug receptors, the active sites are chiral, and they only interact with molecules of specific configuration. This has synthetic chemists became convinced to accept that all compounds used as pharmaceuticals must be in one of their enantiomeric forms. As a consequence, in the 1980s decade, the Food and Drug Administration (FDA) required pharmaceutical industry to acquire drug candidates in details of the toxic effects of the enantiomers. By this, chemical substances to be used as drugs candidates must be synthetized as optically pure compounds or to be highly enriched.

Biologically active chiral molecules have been extracted from natural products has plants. Extracts from the *Ephedra* sp. genus have been traditionally used in Chinese medicine as nasal descongestives, cardiac stimulant, and antiasthma agents. The active principle from this plant was first extracted in 1885, isolated, and then purified in 1887 by Nagai [1] who called it ephedrine. The herb, "Ma-Huang" is the best source of ephedrine, up to 1% bulk weight has been obtained from this material. Amounts of *pseudo*ephedrine, N-methylephedrine, N-methyl*pseudo*ephedrine, *nor*ephedrine, and *norpseudo*ephedrine were found from this herb [2]. Since pharmacologycal studies done by Chen and Schmidt in 1924 [3], the chemists have been interested in the synthesis of physiologically active analogous of ephedrine derivatives [4]. At the present time, large quantities are used in Western medicine to relieve mucous membrane congestion [5].

Today, ephedrine is a pharmaceutical classified as sympathomimetic agent, weaker but longer acting than adrenaline. It acts as cardiac stimulant, hypertensive agent, hyperglycaemic, and bronchodilator. Ephedrine has been clinically used against hay feber, bronchial asthma, myasthenia gravis, whooping cough, Heart block (Stokes-Adam syndrome), and dysmenorrhea. Because ephedrine crosses the hematoencephalic and placentary barriers, have effects on the central nervous system, in consequence, decrease fatigue, sleep (insomnia), and hungry sensations (anorexia) [4].

The no-polar structure of ephedrine makes this substance more liposoluble than catecholamines. It is thermodynamically more stable, in consequence, it is not a substrate for monoamineoxidase (MAO) or the catechol-O-methyltranspherase. Thus, it is a diffusible pharmaceutical that has a more prolonged effect as catecholamines [6].

On the other hand, the modern chemistry is interested in the development of new synthetic methods to produce drugs, antibiotics, alimentary additives, etc., with high optical purity. Asymmetric synthesis design requires catalysts, chiral auxiliaries, and reagents able to control the stereochemistry of the reaction products and to be efficiently recycled [7]. Ephedrines, *nor*ephedrines and its derivatives, have been broadly used as chiral auxiliaries in asymmetric synthesis [8]. Thiols, sulfides, and disulfides obtained from ephedrines have been proven to be very good chiral catalysts [9]. It has been found the use of polymer-supported catalysts applied to organic synthesis with emphasis given to the use of ephedrine chiral catalyst to promote asymmetric reactions [10].

2. Structure of ephedrines

The structure of ephedrine and *pseudo*ephedrine was studied by Ladenburg and Oelschägel. They suggested the formula $PhCH(OH)CH(CH_3)NHCH_3$ now accepted for the alkaloids [11]. Studies that support this formula were provided by Schmith and Bümming [12]. The more important fact that ephedrine and *pseudo*ephedrine are stereoisomers is easy with which ephedrine can be isomerized to *pseudo*ephedrine by acylation or by boiling with HCl (25%) [13], this change has been found to be reversible [14, 15].

Freudenberg and Leithe investigated the configuration about C1 and C2 for ephedrine and *pseudo*ephedrine [16–18], and represented the distribution about these centers of asymmetry, for ephedrine by structure **1a** and for *pseudo*ephedrine by structure **2a**, **Figure 1**.

http://dx.doi.org/10.5772/67387

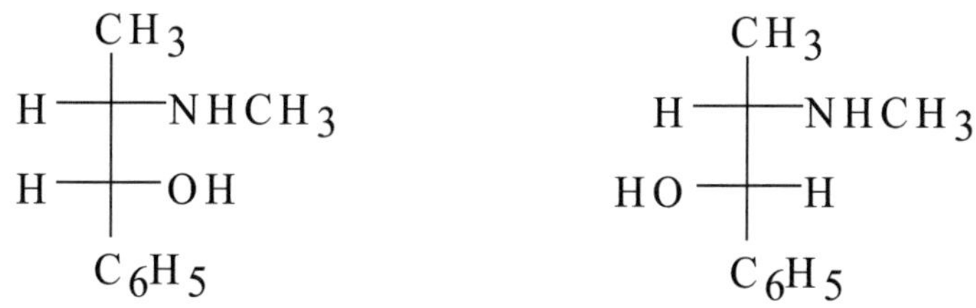

1a (-)-ephedrine **2a** (+)-*pseudo*ephedrine

Figure 1. Freudenberg and Leithe representation of ephedrine **1a** and *pseudo*ephedrine **2a**.

Zhu et al. analyzed the relationship of the substituents of the stereogenic center and to the specific optical rotation. The variables used as matrix elements include (1) the substituent masses (**m**), (2) radii (**r**), (3) symmetries (**s**), and (4) electronegativities (χ) of the atoms or groups bounded to the stereogenic center. For ephedrine and *pseudo*ephedrine, the calculated values were approximate to the observed rotation values [19]. The preferred conformation of ephedrine **1** and *pseudo*ephedrine **2** was theme of controversy [20–25]. The questions could be answered by the X-ray technique.

Several crystal structures of ephedrine salts were reported: the hydrochloride by Bergin [26] and the hydrogen and the di-hydrogen phosphates by Bugg [27, 28] showed the conformation **1b**. On the other hand, in an X-ray study, Mathew et al. demonstrated the conformation **2b** in structures of (+)-*pseudo*ephedrine and (+)-*pseudo*ephedrine hydrochloride [29]. There was found one strong intermolecular hydrogen bond OH···N in *pseudo*ephedrine which links the molecules into infinite chains around the screw axis. An intramolecular contact N—H---O was observed, but the angle of 108° is not favorable. On the other hand, the C(2)—N bond is nearly parallel to the C(1)—C(*ipso*) bond. This conformation was also found in a bis-(+)-*pseudo*ephedrine complex of Koper II [30], (−)-*nor*adrenaline [31], and dopamine [32]. Similar conformations for *nor*ephedrines were found [33] but the difference in energy levels between the various possible conformations in the *nor*-series is less than in the ephedrine series, and interconversion is carried out with ease. This was explained because of the hindering effect of the N-methyl group, **Figure 2** [34].

Finally, ephedrines **1,2** and *nor*ephedrines **3,4** have the β-aminoalcohol structure where the phenyl and methyl groups create two chiral centers on each carbon atom and generate four optically active stereoisomers, **Figures 4** and **5**. Freudenberg et al. [16] and Leithe [17] established

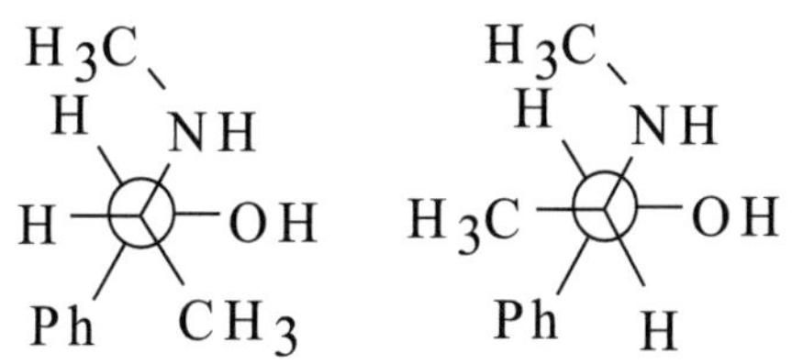

1b **2b**

Figure 2. Stable conformations of ephedrine **1b** and *pseudo*ephedrine **2b**.

the relative configuration about the asymmetric centers for (−)-ephedrine and its optical isomers, hence the configurational relationship between ephedrines **1a,b** and *pseudo*-ephedrines **2a,b** series is well known. The stereoisomers with methyl on nitrogen atom are *l*-(1*R*,2*S*)-ephedrine **1a** and *d*-(1*S*,2*R*)-ephedrine **1b**; *d*-(1*S*,2*S*)-*pseudo*ephedrine **2a,** and *l*-(1*R*,2*R*)-*pseudo*ephedrine **2b, Figure 3**.

The *l*-ephedrine **1a** is the stereoisomer that produces a more pronounced stimulus on the central nervous system, compared with other drugs [35].

The stereoisomers without methyl on nitrogen atom are called *nor*ephedrines: *l*-(1*R*,2*S*)-*nor*ephedrine **3a** and *d*-(1*S*,2*R*)-*nor*ephedrine **3b**; *d*-(1*S*,2*S*) *norpseudo*ephedrine **4a,** and *l*-(1*R*,2*R*)-*norpseudo*ephedrines **4b**, **Figure 4**.

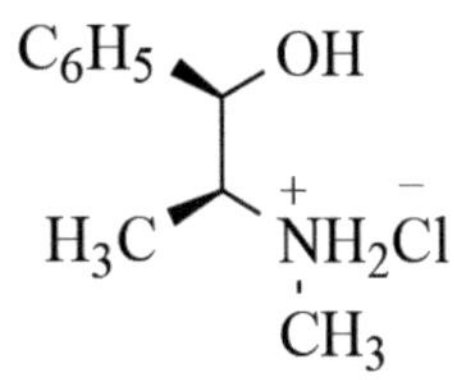

(*1R, 2S*)-(-)-ephedrine
1a

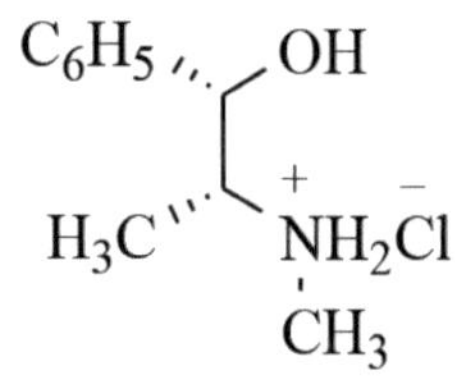

(*1S, 2R*)-(+)-ephedrine
1b

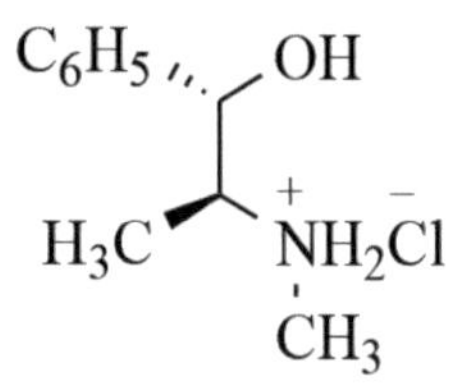

(*1S, 2S*)-(+)-*pseudo*ephedrine
2a

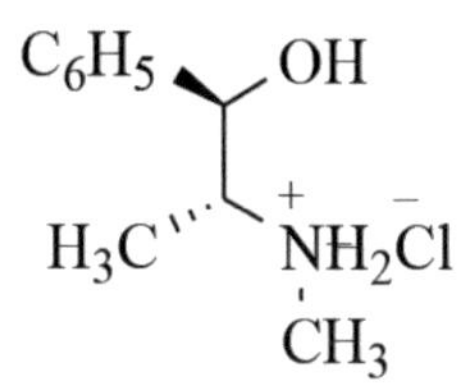

(*1R, 2R*)-(-)-*pseudo*ephedrine
2b

Figure 3. Ephedrine stereoisomers.

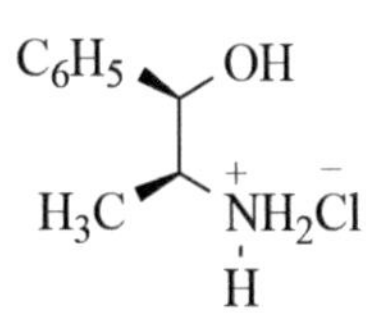

(*1R, 2S*)-(-)-*nor*ephedrine
3a

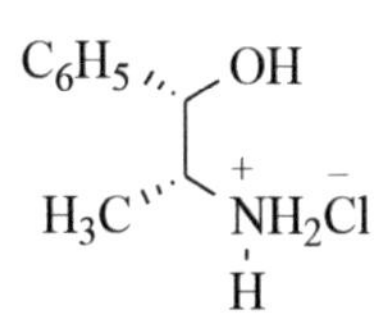

(*1S, 2R*)-(+)-*nor*ephedrine
3b

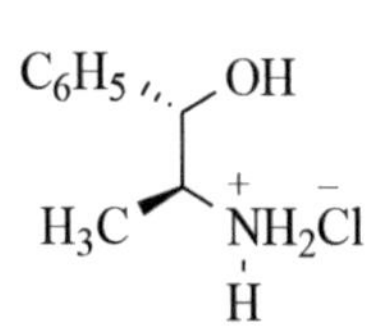

(*1S, 2S*)-(+)-*norpseudo*ephedrine
4a

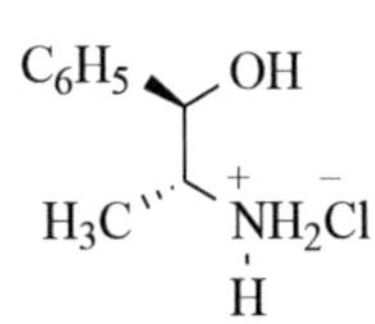

(*1R, 2R*)-(-)-*norpseudo*ephedrine
4b

Figure 4. *Nor*ephedrine stereoisomers.

3. Physical properties of ephedrines

Physical properties of some optical isomers of ephedrines as free bases or as acidic salts have been summarized, **Tables 1–4** [36].

Characteristic	mp (°C)	$[\alpha]_D$
Free base (B)	37–39	–41°
Hemihydrate	39–43	–6.3° (EtOH) +11.2° (H_2O)
Hydrochloride	216–220	–34° (H_2O)
Hydrobromide	205	–
Sulphete	243	–30° (H_2O)
Oxalate	249 (dec)	Insol. H_2O
Aurichloride	128–131	–
Platinichloride	186	–

Table 1. (1*R*, 2*S*)-(–)-ephedrine forms.

Characteristic	mp (°C)	$[\alpha]_D$
Free base (B)	118–120	–51°(EtOH)
Hemihydrate	–	–
Hydrochloride	185–188	–62. (H_2O)
Hydrobromide	–	–
Sulphete	–	–52.5° (H_2O)
Oxalate	218 (dec)	Insol. H_2O
Aurichloride	126.5–127.5	–
Platinilchloride	–	–

Table 2. (1*R*, 2*R*)-(–)-*pseudo*ephedrine forms.

Characteristic	mp (°C)	$[\alpha]_D$ (°)
Free base (B)	51–54	–14.56(EtOH)
Hemihydrate	–	–
Hydrochloride	172–175	–33° (H_2O)
Hydrobromide	–	–
Sulphete	285–286 (dec)	–31.99 (H_2O)
Oxalate	245 (dec)	–
Aurichloride	188	–
Platinichloride	221 (dec.)	–

Table 3. (1*R*, 2*S*)-(–)- *nor*ephedrine forms.

Characteristic	mp (°C)	$[\alpha]_D$
Free base (B)	77–8 (corr.)	−37.9 (MeOH)
Hemihydrate	–	–
Hydrochloride	180–183 (corr.)	−41.7° (H_2O)
Hydrobromide	–	–
Sulphete	295 (dec.)	−48.7° (H_2O)
Oxalate	235 (dec.)	–
Aurichloride	137–138	–
Platinilchloride	198	–

Table 4. (1*R*, 2*R*)-(−)- *norpseudo*ephedrine forms.

4. Biological activity of ephedra heterocycles

A wide approach for the synthesis of new compounds that possess some kind of biological activity is the cyclization of substituted phenethylamines as ephedrines into heterocycles, such as morpholine (phenmetrazine) [37] and 2-amine-oxazolines (4-methylaminorex and 3,4-dimethylaminorex) [38], in such a way that the ephedrine skeleton becomes part of the heterocyclic ring, **Figure 5**.

Some other heterocycles as oxazolidine [39], di- and tetrahidro-1,3,4-oxadiazines [40, 41], 2-thiazoline [42], thiazolidine [43], dihidro-1,3,4-thiadiazine [44], tetrahydrotriazine [45], and imidazolidine [46] derived from ephedrines and *nor*ephedrines have been reported. Certain of these heterocycles exhibit different biological effects as central nervous system, stimulating, appetite-depressing [37, 38], monoamine oxidase inhibiting antidepressant [40g, 44b], central nervous system depressant [40e, f, 41, 45], analgetic [45], hypocholesterolemic [41], anti-inflammatory [41], antimicrobial [41, 44b], or catecholamine-potentiating [43] activities. On the other hand, 3,4-dimethyl-5-phenyl-oxazolidine is used as a prodrug [47].

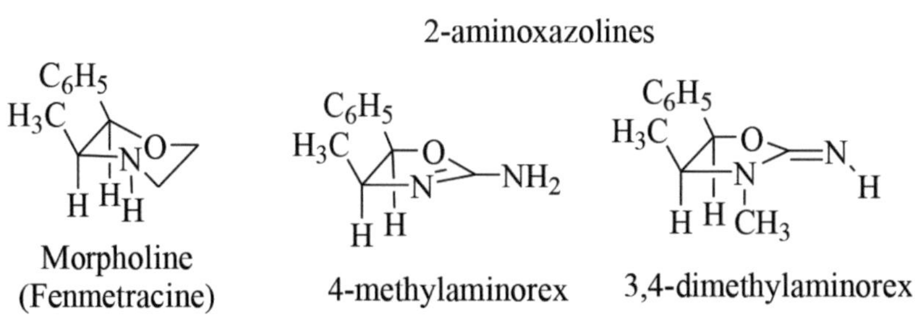

Figure 5. Some heterocyclic compounds with biological activity.

http://dx.doi.org/10.5772/67387

5. Reactions to get chlorodeoxyephedrine derivatives

The bromination reaction of ephedrine hydrochloride **2a** with PBr_5 produce the bromodeoxy-derivative [48]. On the other hand, ephedrine **1a** or *nor*ephedrine **3a** or its hydrochlorides reacts with $SOCl_2$ to give chlorodeoxy*pseudo*ephedrine **5a** or chlorodeoxy*norpseudo*ephedrine **5b**. The same reaction with *pseudo*ephedrines **2a** or **4a** give a 60:40 diastereomeric mixture of *threo:erytho* **5a:6a**, **Scheme 1** [49]. In order to improve the stereoselectivity, chlorination reaction of *pseudo*ephedrine stereoisomers **2** at 0°C was carried out. In these conditions, only the corresponding *threo* chlorodeoxystereoisomers **3** were stereoselectively obtained (S_Ni mechanism), **Scheme 1**, the X-ray diffraction structure of **5a** is depicted in **Figure 6**.

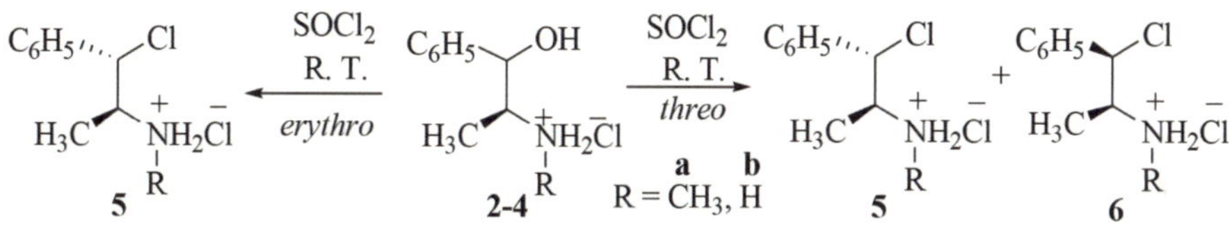

Scheme 1. Chlorodeoxyephedrine hydrochlorides from chlorination reaction of ephedrines **2**.

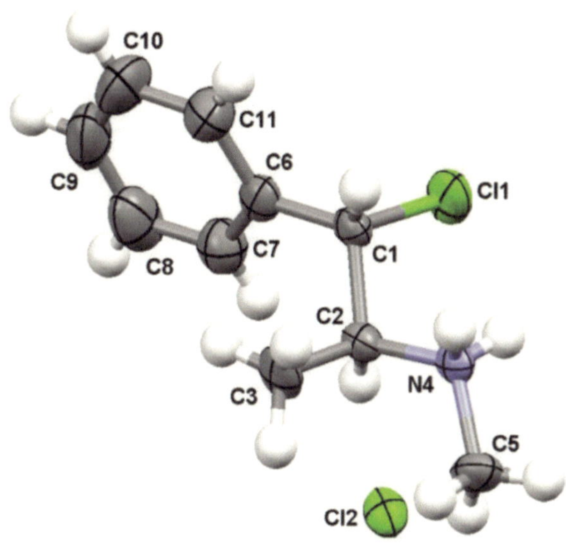

Figure 6. X-ray diffraction structure of Chlorodeoxyephedrine hydrochloride **5a**.

6. Heterazolidines-2-heterounsaturated from ephedrines

In 1950, Close reported the solvent-free dehydration of ephedrine hydrochloride **1** in the presence of urea at 180–200°C to afford the imidazolidinone **11a**-(*t*) and oxazolidinone **15a**-(*c*),

Scheme 2 [34]. In that work, it was proposed that urea is converted into ammonium oxocyanate at 180–200°C, which in the presence of hydrogen chloride, ammonium chloride and oxocyanic acid are produced. Finally, ephedrine reacts with oxocyanide acid to produce the nonisolated urea intermediate **7a**-(*e*), which cyclization by dehydration affords the *trans*-imidazolidone **11a**-(*t*). On the other hand, in a simultaneous manner, cyclization of urea **7a**-(*e*) by nucleophylic attack of the oxygen atom of ephedrine to the ureidic carbonyl and ammonia elimination produces the oxazolidone heterocycle **15a**-(*c*).

The same reaction was revisited with the use of K^+NCO^- instead of urea, and the study was extended with thiocyanates and *pseudo*ephedrine **2a**-(*th*), *nor*ephedrine **3b**-(*e*), and *norpseudo*ephedrine **4b**-(*th*), as stereoisomers to produce a series of optically active 1,3-heteroazolidine-2-heterounsaturated compounds **9–15, Scheme 3.**

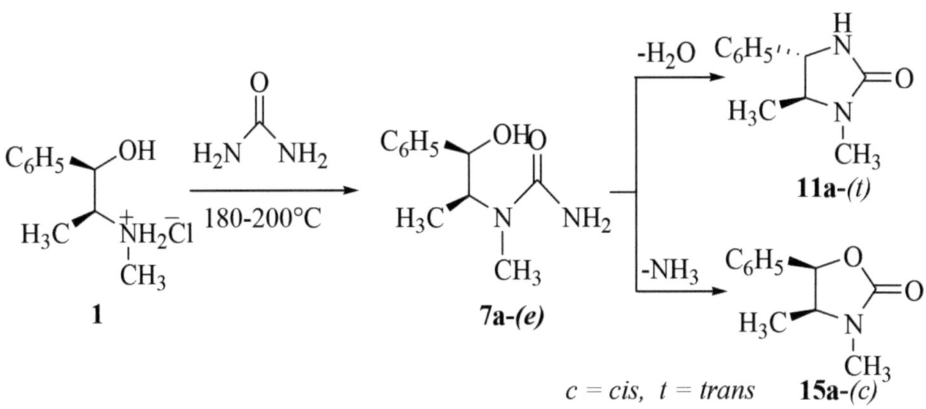

Scheme 2. Dehydration of ephedrine 1a-(*e*) with urea according to Close.

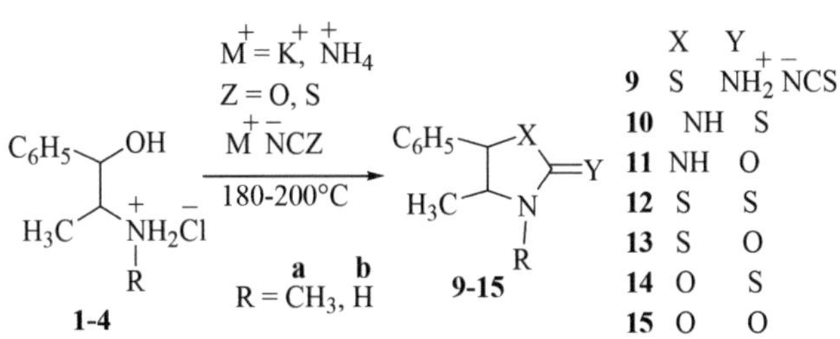

Scheme 3. 1,3-heterazolidines-2-heterounsaturated from ephedrines.

6.1. Reaction of ephedrines with oxocyanate

Urea intermediate **7a**-(*e*), proposed in the Close's reaction, was isolated when K^+NCO^- is reacted with ephedrine hydrochloride **1** in refluxing ethanol for 72 h (78% yield). The reaction was also performed with *pseudo*ephedrine **2**, *nor*ephedrine **3**, and *norpseudo*ephedrine to afford, the urea derivatives **7a**-(*th*), **7b**-(*e*), and **7b**-(*th*) in 83, 80, and 86% yield, respectively

(**Scheme 4**). In the case of the reaction of $Na^{+}NCS^{-}$ with ephedrine stereoisomers series **1–4**, only chloride by thiocyanate anion was exchanged to give the corresponding hydrothiocyanates **8a,b** (*e,th*) (Scheme 4). The urea intermediate **7a**-(*e*) derived from ephedrine could be crystallized from ethanol and its structure studied by X-ray diffraction (**Figure 7**).

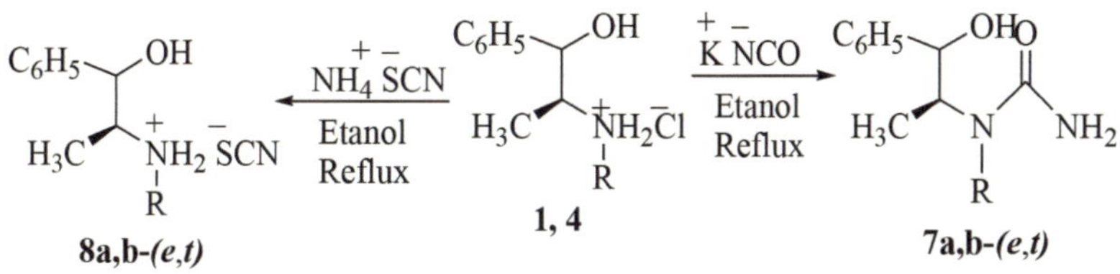

Scheme 4. Reaction of $K^{+}NCO^{-}$ and $NH_4^{+}NCS^{-}$ with ephedrines 1a,b-(*e,th*) in refluxing ethanol.

An intramolecular hydrogen bonding interaction between the hydrogen atom of the hydroxyl group and the ureidic oxygen atom to form a seven membered ring was observed. The O1H1···O6 distance of 1.820(24) Å [angle of 166.72° (2.25)] represents a strong interaction [50]. The formed hydrogen bond forces the NH_2 group to adopt a syn conformation to the N-Me group [C8N4C5N7 angle of −6.00 (0.26)°]. In addition, both N—CO bond distances are of intermediate value between a single (1.469 Å) and a double (1.279) N—C bond (1.35 Å mean) [51].

When the intermediate **7a**-(*e*) was free of solvent heated at 180–200°C for 1 hour, an equimolar mixture of imidazolidone **11a**-(*c*) and oxazolidone **15a**-(*c*) was formed (**Scheme 5**). Imidazolidinone **11a**-(*c*) was separated by precipitation from a $CHCl_3$ and purified by recrystallization from ethanol. The structure of *cis* (*c*) isomer instead of the expected *trans* (*t*) isomer [52] was observed on the ^{1}H and ^{13}C NMR spectra and confirmed by X-ray diffraction analysis. The formation of an aziridinim isocyanate **I** then the isocyanate **II** as intermediates

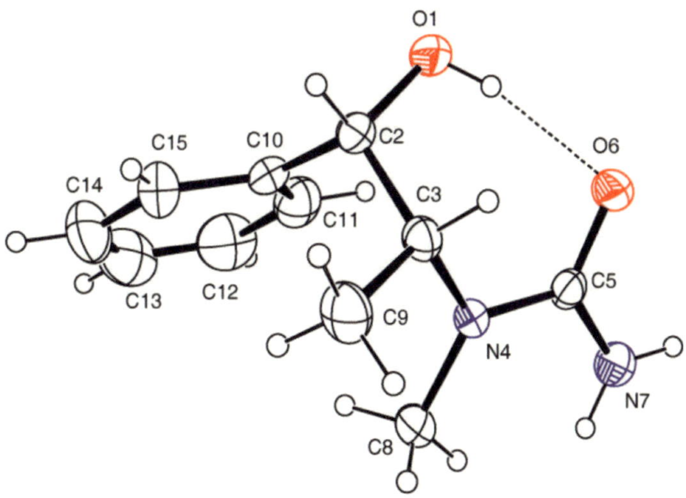

Figure 7. Molecular structure of ephedrine-urea **7a**-(*e*).

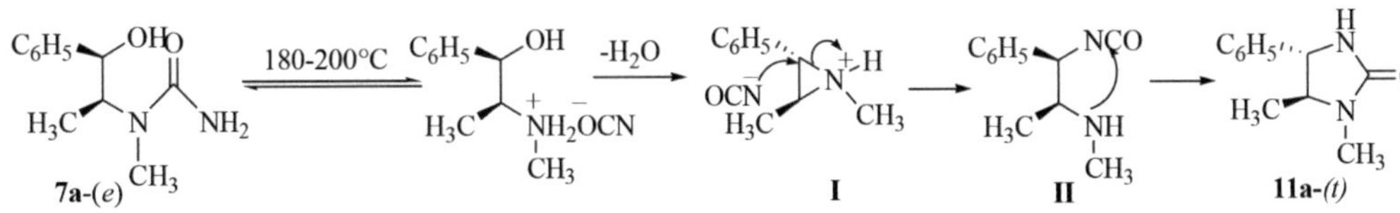

Scheme 5. Mechanistic pat way for the cyclization of ephedrine-urea **7a**-(*e*) to get imidazolidinone **11a**-(t).

are proposed to explain the retention of C1 configuration in the formation of the *cis*-imidazolidone **11a**-(*c*) (**Scheme 5**) [53]. On the other hand, the oxazolidone **15a**-(*c*) is formed in accord to the Close's idea (**Scheme 2**). The *in situ* formation of amides from aminoalcohols involved in oxazolidine formation has been reported in the literature [34, 54].

On the basis of these previous findings, the free of solvent reaction of ephedrines **1,4** with sodium or ammonium thiocyanates were performed.

6.2. Reaction of ephedrines with thiocyanate

In the direct heating of one equimolar of $Na^{+}NCS^{-}$ with ephedrine 1 at 180–200°C during 0.5 h, ephedrine hydrothiocyanate 8a-(*e*) from the aqueous phase and *trans*-thiazolidine-2-imino hydrothiocyanate 9a-(*t*) from the chloroform phase (10% yield) were separated, after $CHCl_3/H_2O$ partition, **Scheme 6**. Deamination of ephedrine hydrothiocyanate 8a-(*e*) proceeded to give Ethylphenylketone as lateral product.

The reaction with two molar equivalents of NH_4SCN for 4 hours afforded the thiazolidine-2-imino hydrothiocyanate **9a**-(*t*) in 50% yield as precipitate from $CHCl_3$. The use of NH_4SCN instead of NaSCN salt, avoided deamination, due to its lower melting point (153°C). Compound **9a**-(*t*) was identified on comparing the structure obtained from chlorodeoxy*pseudo*ephedrine [55]. On the other hand, several heterocycles were separated when the remaining chloroform mixture was eluted in a chromatographic column. The mass spectrometry of the separated fractions showed the presence of heterocycles summarized in **Table 5**.

As described above for ephedrine **1**, two molar equivalents of $NH_4^{+}NCS^{-}$ were heated with *pseudo*ephedrine **2**, *nor*ephedrine **3**, and *norpseudo*ephedrine **4**. The identified compounds are listed in **Table 5**.

The reaction mixture of *pseudo*ephedrine hydrochloride **2** was treated with a 50:50 of $CHCl_3/H_2O$ mixture. The chloroform phase was eluted in a column chromatography. Using chloroform as eluent, the imidazolidine-2-thione **10a**-(*c*) was separated in 40% yield as first fraction.

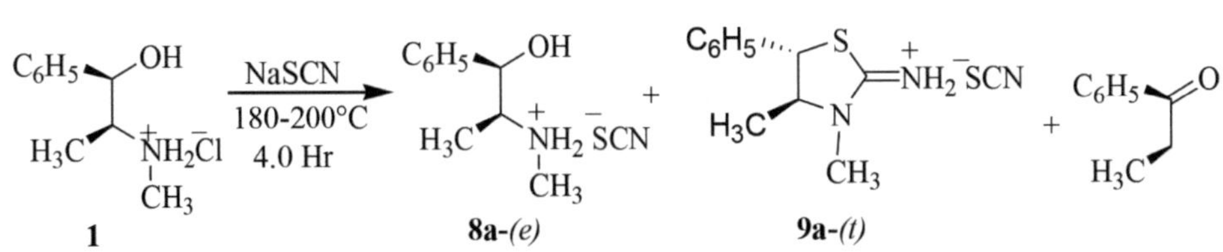

Scheme 6. Heating reaction of ephedrine **1** with NaSCN.

	Heterocycle	**9**	**10**	**11**	**12**	**13**	**14**	**15**
Ephedrine	X	S	NH	NH	S	S	O	O
	Y	NH_2SCN	S	O	S	O	S	O
1a-(*e*)	*Cis*		184 (2)	163 (3) [190(27)]	196 (7) [223 (100)]	172 (4)		160 (2) [191 (19)]
	Trans	168(**50**) [206(98)]	183 (7)	162 (7)	195 (2) [223 (100)]	171(7)		191 (23)
1a-(*th*)	*Cis*	169(15) [206(15)]	184 (**40**) [206 (100)]	163 (5) [190(27)]	196 (6) [223 (100)]	172 [207(64)]		160 (5) [191 (19)]
	Trans			162 (2)	Traces [223 (100)]	171 [207(41)]		159 (**20**) [191 (23)]
1b-(*e*)	*Cis*		Traces [192 (100)]		200 (5) [209 (100)]	176 (5) [193(16)]	189 (2) [193 (35)]	160 (**40**) [177 (8)]
	Trans		Traces		199 (2) [209 (100)]	175(**40**) [193(25)]		
1b-(*th*)	*Cis*	173(**40**) [192(62)]	183 (10) [192 (100)]		200 (15) [209 (100)]	175(10) [193(16)]		160 (3)
	Tans	172(10)				174 (5)		159 (3)

Table 5. Carbonyl carbon chemical shift in ppm, proportion (%) and mass spectrometry data [M^+ (%)] of heterocycles **9–15**.

Imidazolidones **11a**-(*c*) and **11a**-(*t*), thiazolidinedione **12a**-(*c*), and thiazolidinones **13a**-(*c*) and **13a**-(*t*) were separated in small quantities in the subsequent three remaining fractions, respectively. On the other hand, after evaporation of the aqueous phase, thiazolidine-2-imino hydrothiocyanate **9a**-(*c*) and oxazolidinones **15a**-(*c*) and **15a**-(*t*) as solid mixtures were identified by mass spectrometry. NMR spectral data of compound **10a**-(*c*) show a broad signal at 6.32 ppm (^1H) and at 183 ppm (^{13}C) of the N—H and C—S groups, respectively. The molecular ion [z/e = 206 (100%), M^+] and X-ray diffraction analysis confirmed the structure, **Figure 8**. The bond distances C2—N1 [1.345(14) Å] and C2—N3 [1.332(13) Å] show an intermediate value between a single and a double bond [51], due to a conjugation through the N1—C2—N3 fragment.

From the reaction mixture of *nor*ephedrine hydrochloride **3**, compounds **13b**-(*c*) and **13b**-(*t*) in a 1:8 proportion were separated by chromatography. Compound **13b**-(*t*) (40% yield) was separated from **13b**-(*c*) in a second column using $CHCl_3$. The spectra data, molecular ion [z/e = 193 (25%), M^+], and the X-ray diffraction structures of compound **13b**-(*t*) confirm the *trans* configuration (**Figure 9**). A conjugation through the N—C2—S fragment is observed because the bond distances C2-N [1.332(3) Å] and C2-S [1.775(2) Å] are shorter than the corresponding single bonds.

The reaction mixture of *norpseudoe*phedrine hydrochloride **4** was separated in a chromatographic column and each fraction analyzed by mass spectrometry, **Table 5**. The fourth fraction contained thiazoline-2-amine hydrothiocyanate **9b**-(*c*).

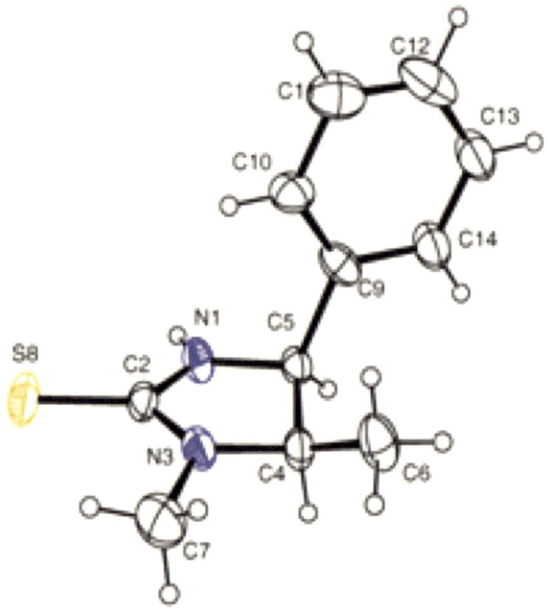

Figure 8. Structure of imidazolidinethione **10a**-(*c*).

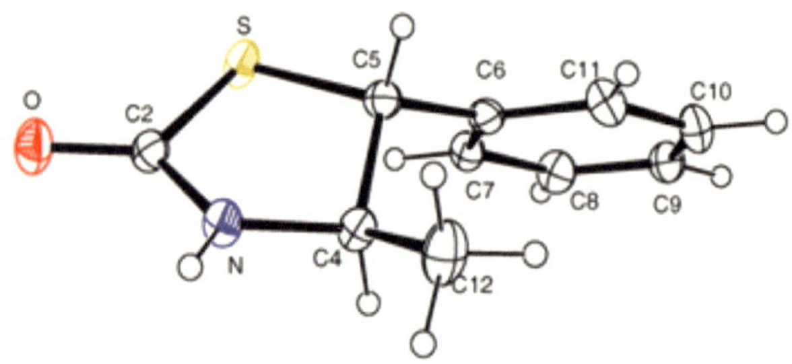

Figure 9. Structure of***trans***-thiazolidinone **13b**-(*t*).

At least four competitive mechanisms are proposed to explain the formation of heterocycles **9-15a,b** in the heating reaction of NH_4SCN with ephedrines **1,2** and *nor*ephedrines **3,4** (**Scheme 7**). In general, with exception of heating reaction of *nor*ephedrine **3**, a S_N2 dehydration mechanism by the thiocyanate = isothiocyanate anions as nucleophiles and the subsequent cyclization of the ephedrinethiocyanate (**IV**) and/or ephedrineisothiocyanate (**III**) intermediates formed operate to give the corresponding thiazolidine-2-imino hydrothiocyanates **9a,b** and/ or imidazolidinetiones **10a,b**. The product from thiocyanate predominate in the heating reaction of ephedrine **1**, and for *pseudo*ephedrine **2**, the product from isothiocyanate predominate. A stable alkyl ephedrinethiocynate analogue to **IV** has been isolated, which support the proposed mechanism [56].

In the heating reaction of *nor*ephedrine **3**, the H_2S obtained by hydrolysis of thiocyanate acts as nucleophyle in competitive S_N1 and S_N2 mechanisms to form the thiolephedrine thiourea

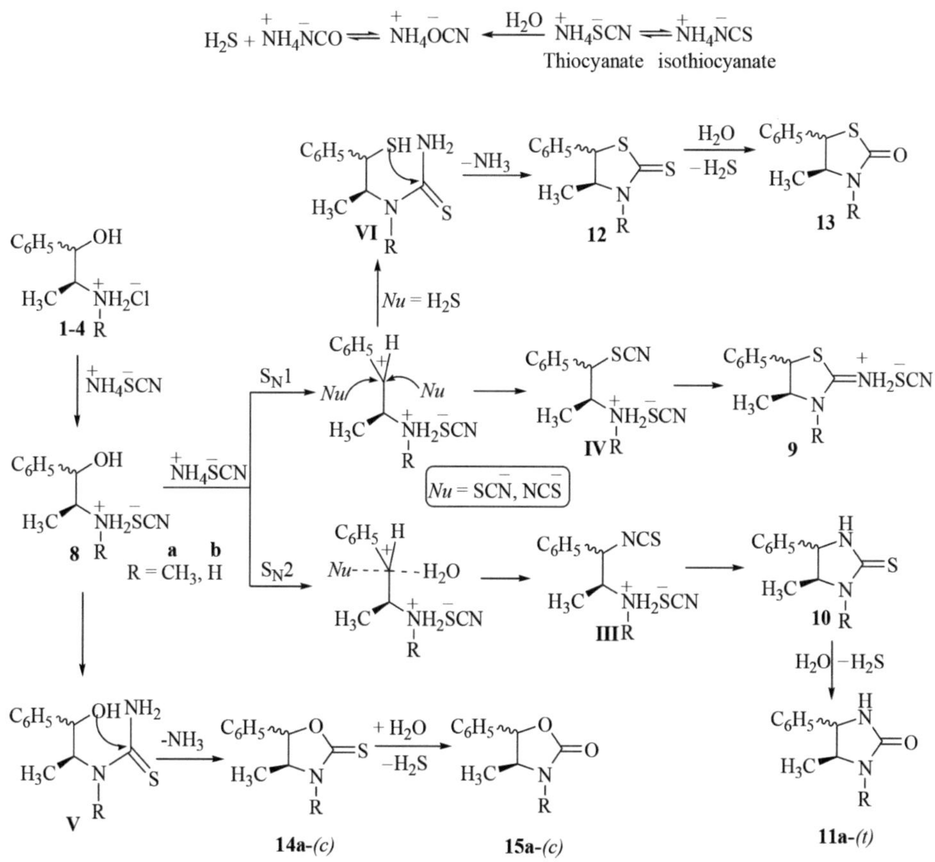

Scheme 7. Proposed mechanisms to explain the formation of heterocycles **9–15**.

VI-(*e,th*), which cyclization afford thiazolidinethiones **12b**(*c,t*). A desulphurization by hydrolysis of thiazolidinethiones **12b**-(*c,t*) explains the formation of thiazolidinones **13b**-(*c,t*). A mechanism through thioureidic intermediate **V** operates simultaneously, its cyclization affords oxazolidinethione **14b**-(*c*), which desulphurization gives oxazolidinone **15b**-(*c*).

Desulphurization of the oxazolidinethione **14a**-(*t*) explains the formation of oxazolidinone **15a**-(*t*) (20%) in the heating reaction of *pseudo*ephedrine **2**, **Scheme 6**, this mechanism is favored when one molar equivalent of $NH_4^+NCS^-$ is used. In this case, oxazolidinone **15a**-(*t*) (40%) and imidazolidinethione **10a**-(*c*) (20%) were obtained as the major products. Similar results were observed in the heating reaction of *norpseudo*ephedrine **4**. If $NH_4^+NCS^-$ is changed from two to one molar equivalents, thiazoline-2-amine hydrothiocyanates **9b**-(*c*) decreased from 40 to 15% and oxazolidinone **15b**-(*t*) increased from 3 to 45%.

In general, in the ^1H NMR spectral data of ephedracycles, the $C-CH_3$ group of the *cis* isomers appears at low frequency shifts in the range between 0.9 and 0.7 ppm, compared with the same group of the *trans* isomers, appearing between 1.1 and 1.4 ppm, this is due to the shielding effect of the phenyl group.

7. Heterazolidines-2-heteroinsaturated from chloro*pseudo* ephedrines

In continuation with our investigations on the design of new heterocycles derived from ephedrines **1**, in this work, we revisited the cyclization reactions of chlorodeoxy*pseudo*-ephedrine hydrochloride **5a**-(*th*) (R = Me) with one or two molar equivalents of potassium oxocyanate, sodium thiocyanate, and potassium selenocyanate nucleophiles as cyclizing agents in refluxing ethanol. In addition, the results of the reaction of chlorodeoxy*nor*p-seudoephedrine hydrochloride **5b**-(*th*) (R = H) with the above mentioned nucleophiles are reported. An interesting finding of this study was the synthesis of the *trans* isomer of 1,3-oxazolidine-2-iminium chloride **18a**-(*t*) through the *in situ* chlorinated urea intermediate **7a**-(*e*), **Scheme 8**.

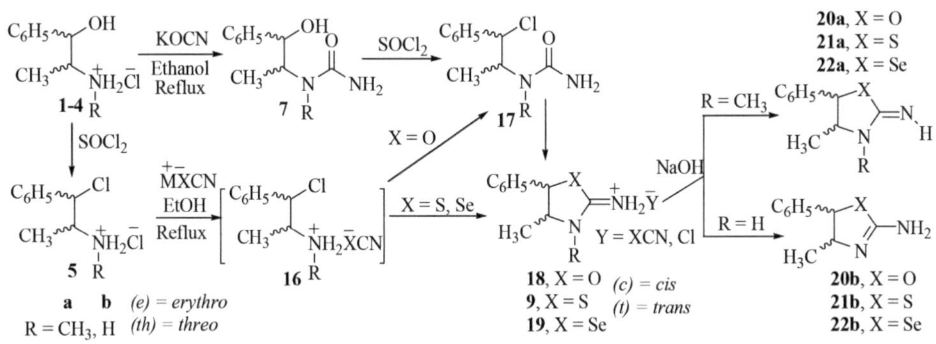

Scheme 8. Reactivity of chlorodeoxy*pseudo*ephedrine hydrochlorides **5** with heterocyanates.

7.1. Reaction of chlorodeoxy*pseudo*ephedrine hydrochlorides 5 with potassium oxocyanate

Chlorodeoxy*pseudo*ephedrine hydrochloride **5a**-(*th*) was reacted with two molar equivalents of KOCN in stirring ethanol at room temperature. The reaction was monitored at 24, 48, and 72 h by ^{1}H NMR. Two compounds, in 80:20, 60:40, and 40:60 proportions, respectively, were observed. The NMR tube of the 40:60 proportion in DMSO-d_6 was heated at 92°C during 1 h, to quantitatively transform the minor proportion compound into the 1,3-oxazolidine-2-iminium oxocyanate **18a**-(*c*) identified as the only product. The *N*-(1-chloro-1-phenyl-2-methyl-ethyl)-*N*-methyl urea **17a**-(*th*) in the 80:20 mixture with **18a**-(*c*) was identified as the intermediate. The use of one molar equivalent of potassium oxocyanate in the same reaction in refluxing 16 h afforded the hydrochloride of the oxazolidine-2-imine **18a**-(*c*), which was crystallized from ethanol to be analyzed by X-ray diffraction, the structure is shown in the **Figure 10**.

The reaction is general; the reaction of one molar equivalent of KOCN with chlorodeoxy*norp-seudo*ephedrine hydrochloride **5b**-(*th*) (R = H) in refluxing ethanol 8 h afforded the chlorourea derivative **17b**-(*th*). The proposed mechanistic pathway represented in **Scheme 9** explains why the reaction is carried out with inversion of the Cl configuration to get the *cis* isomer.

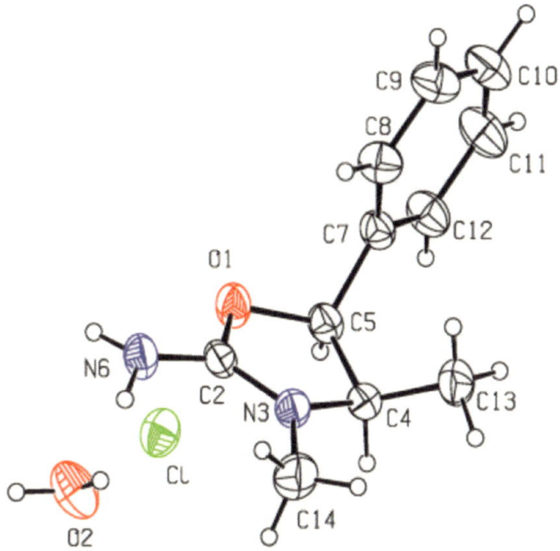

Figure 10. Molecular structure of hydrochloride compound **18a**-(*c*).

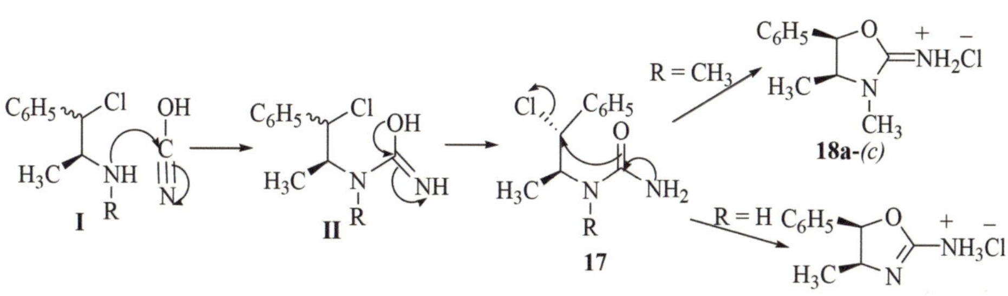

Scheme 9. Mechanistic pathway involved in the synthesis of compounds **18a**-(*c*) and **18b**-(*c*).

Chlorourea compound **17b**-(*th*) could be isolated and characterized by NMR. Two signals are observed in the ^1H NMR spectrum 6.04, (d, $^3J = 8.5$ Hz) and 5.55 ppm (s, broad) in a 1:2 proportion, respectively, assigned to NH and NH_2 urea hydrogen atoms, respectively. The NH coupling constant value proposes a hydrogen bonding NH---Cl interaction, which makes this hydrogen and H2 to be in an *anti*position. In addition, the small H^1, H^2 coupling constant ($^3J = 5.28$ Hz) supports this proposed interaction, **Figure 11**. The ^{13}C NMR spectrum shows the carbonyl carbon signal at 159.6 ppm, according to the proposed structure.

The chlorourea derivative **17b**-(*th*) was refluxed in ethanol during 24 h. The ^1H NMR spectrum of the solid precipitated showed a 80:20 mixture of two heterocycles. The ^1H NMR chemical shift of the $^+NH_3$ group appears in 9.65 ppm as a broad signal, H5 and H4 appear at 6.49 (d) and 4.56 ppm (dq), for the major compound. For the minor compound, H5 and H4 appear at 5.26 (d) and 5.41ppm (dq), respectively. In both compounds, the

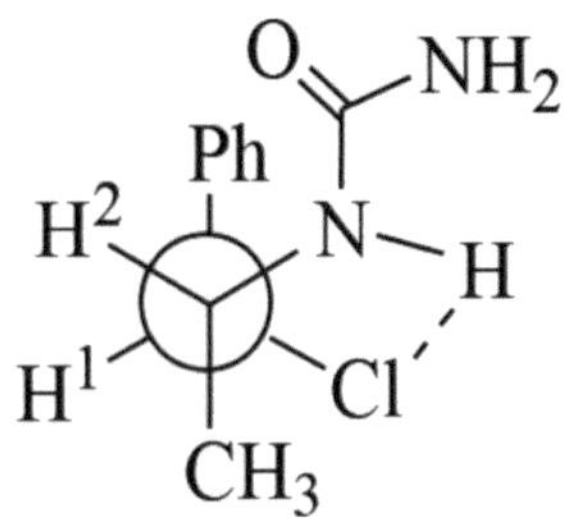

Figure 11. Hydrogen bonding interaction proposed in compound **17b**-(*th*).

coupling constants are of the same value. In addition, the multiplicity of these signals are interchanged for the minor compound, which are correlated with ^{13}C NMR signals at 65.9 (C4) and 84.6 (C5) ppm, respectively. These results allowed us to assign the *cis*-4-methyl-5-phenyloxazoline-2-ammonium chloride **18b**-(*c*) as the major compound the *cis*-5-methyl-4-phenyl-oxazoline-2-ammonium hydrochloride **23b**-(*c*), as the minor compound, whose formation is explained due to the participation of an aziridine intermediate III, **Scheme 10** [52].

It is known that in chlorination reaction of ephedrine derivatives with thionyl chloride, the C1 configuration is retained through a S_Ni mechanism, when ephedrine bears a bulky group as oxamide or sulfonamide on the nitrogen atom [57]. In this sense, we obtained the *erythro* isomer of ephedrineurea intermediate **7a**-(*e*) by the reaction of ephedrine hydrochloride **1a**-(*e*) with KOCN [58]. This ephedrineurea was chlorinated with thionyl chloride in $CHCl_3$ to get, *in situ*, 1-(2-chloro-1-methyl-2-phenyl-ethyl)-1-methyl-urea **17a**-(*e*). Compound **17a**-(*e*) was refluxed in ethanol during 8 h. ^1H and ^{13}C NMR spectroscopic data of the solid obtained after solvent removal allowed us to identify the *trans* isomer of 3,4-dimethyl-5-phenyl-oxazolidine-2-iminium chloride **18a**-(*t*). This result showed that chlorodeoxyephedrine urea **17a**-(*e*) was obtained with retention of C1 configuration, which was cyclized with the inversion of C1 configuration to obtain **18a**-(*t*). In a similar

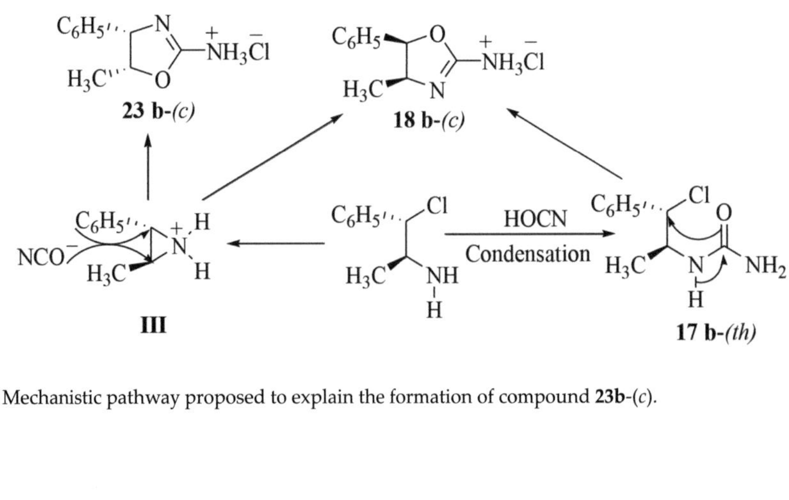

Scheme 10. Mechanistic pathway proposed to explain the formation of compound **23b**-(*c*).

manner, the same reaction with *nor*ephedrine hydrochloride **1b**-(*e*) is stereoselective to get the *cis* isomer of the oxazoline-2-ammonium chloride **18b**-(*c*). In contrast, the same procedure for *pseudo*ephedrine **1a**-(*th*) and *norpseudo*ephedrine **1b**-(*th*) hydrochlorides gave a mixture of oxazolidine-2-iminium chlorides **18a** (60:40, *c:t*) and oxazoline-2-ammonium chlorides **18b** (75:25, *c:t*), respectively [59].

7.2. Reaction of chlorodeoxy*pseudo*ephedrine hydrochlorides 2 with sodium thiocyanate

It is known that the condensation reaction of chlorodeoxy*pseudo*ephedrine hydrochloride **5a**-(*th*) with two molar equivalents of NaSCN in refluxing ethanol for 8 h stereoselectively affords the *trans*-thiazolidine-2-iminio thiocyanate **9a**-(*t*) [51].

The reaction of chlorodeoxy*norpseudo*ephedrine hydrochloride **5b**-(*th*) (R = H) with two molar equivalents of KSCN in refluxing ethanol, only chloride is interchanged by thiocyanate anion to give chlorodeoxy*norpseudo*ephedrine hydrothiocyanate **16b**-(*th*) (ν = 2057 cm^{-1}, ^{-}SCN), even at 24 h of reflux. If hydrothiocyanate **16b**-(*th*) in DMSO-d_6 is heated (90°C) 1 h in a NMR tube, a 50:50 *cis/trans* mixture of 1,3-thiazoline-2-ammonium thiocyanate **9b** was detected in the ^{1}H NMR spectrum. However, only the *cis* isomer of **9b**-(*c*) was stereoselectively produced if the reaction is solvent free heated at 170°C during 3 hours, **Scheme 8**.

7.3. Reaction of chlorodeoxy*pseudo*ephedrine hydrochlorides 2 with potassium selenocyanate

As previous result reported for chlorodeoxy*pseudo*ephedrine hydrochloride **5a**-(*th*) [51], the reaction of chlorodeoxy*norpseudo*ephedrine hydrochloride **5b**-(*th*) with two equivalents of KSeCN in refluxing ethanol for 10 hours affords *trans*-selenazoline-2-ammonium selenocyanate **19b**-(*t*). On the other hand, if only one equivalent of KOCN, NaSCN, or KSeCN is used in the reactions, the corresponding hydrochloride salts of the 2-aminoheterocyles are obtained. Both XCN^- (X = O, S, Se) and Cl^- salts were liberated with aqueous NaOH to give the corresponding imine **20–22a** or amine **20–22b** compounds. Compound **20b**-(*c*) and **22b**-(*t*) were crystallized from ethanol and chloroform, respectively. The structures could be established for X-ray diffraction analysis, **Figures 12** and **13**, respectively.

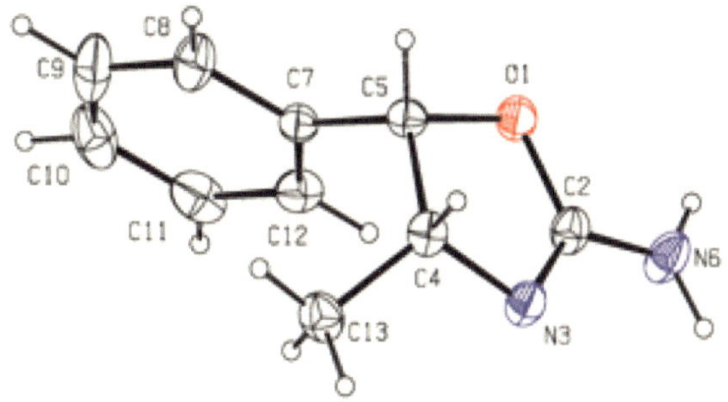

Figure 12. X-ray diffraction structure of **20b**-(*c*).

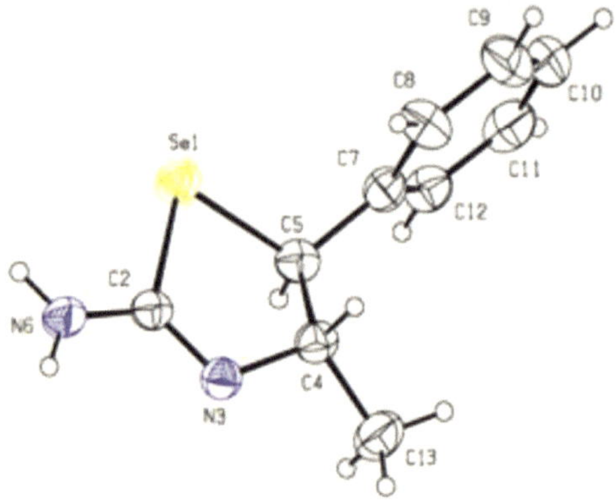

Figure 13. X-ray diffraction structure of **22b**-(*t*).

8. *CIS*-thiazolidinethiones from chloro*pseudo*ephedrines

In 1995, we reported the reaction of chlorodeoxy*pseudo*ephedrine.hydrochloride **5a**-(*th*) with 33% aqueous solution of sodium trithiocarbonate (Na_2CS_3) in refluxing ethanol to give *cis*-thiazolidinethione **12a**-(*c*) (53% yield), **Scheme 11** [60]. However, the same reaction with chlorodeoxy*norpseudo*ephedrine **5b**-(*th*) failed to give the corresponding *cis*-thiazolidinethione derivative **12b**-(*c*).

By this, we encourage us the goal to selectively obtain *cis*- or *trans*-thiazolidinethiones **12a,b** from either chlorodeoxy*norpseudo*ephedrine **5b** or chlorodeoxy*pseudo*ephedrine **5a** derived from ephedrines **1,3**.

To get thiazolidinethiones **12a,b**, the chlorhydrates of chlorodeoxy*pseudo*ephedrines **5a** or **5b** were reacted with one molar equivalent of sodium dithiocarbonate in ethanol solution at room temperature. In the case of chlorodeoxy*norpseudo*ephedrine·HCl **5b**, a white powder solid was precipitated in stirring ethanol for 6 h. The *cis* and *trans* relationships between the phenyl and the methyl groups in thiazolidinethiones **12b** was deduced from the analysis of their ^{1}H and ^{13}C NMR spectral data and are in agreement with data reported [49d]. On this bases, the product represented a mixture of *cis*:*trans*-thiazolidinethiones of **8** in a 9:1 proportion. A S_N2 mechanism to explain the C1 inversion of configuration, then cyclization to get the *cis*-isomer is proposed to be carried out, **Scheme 11**. In addition, a competitive double S_N2 mechanism on C1, then cyclization in which *cis*-aziridine **24b**-(*c*) as intermediate is involved to explain the presence of the *trans*-isomer **12b**-(*t*). Analogous mechanistic observations were proposed

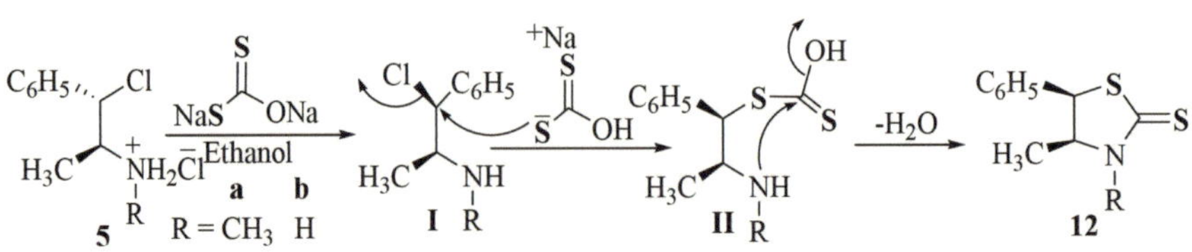

Scheme 11. Mechanistic transformation to get *cis*-thiazolidinethiones from chlorodeoxy*pseudo*ephedrines.

to get stereospecifically thiazolidinethiones from the reaction of viciodoalkanecarbamates with potassium ethylxanthate [61]. When the same reaction at 0°C for 6 h was performed, only *cis*-thiazolidinethione **12b-**(*c*) was precipitated as a white powder in 95% yield. Thiazolidinethione **12b-**(*c*) is stable as thione tautomer in concentrated solution (*δ* NH at 8.3 ppm). However, in a diluted solution, the thiol tautomer is present (*δ* SH at 1.6 ppm).

The reaction of chlorodeoxy*pseudo*ephedrine·HCl **5a** at 0°C was performed, and after 3 days off, white orthorhombic crystals of *cis*-thiazolidinethione **12a**-(*c*) precipitated in 81% yield. The X-ray diffraction structure showed the *cis*-isomer.

To confirm that the *cis*-aziridine is responsible of the *trans*-thiazolidinethione formation, the Kelloggs method was used with the previously obtained *cis*-aziridine **24a**-(*c*) and **24b**-(*c*) [62] from chlorodeoxy*pseudo*ephedrines **5a** or **5b**. The corresponding *cis*-aziridine was reacted with CS_2 in stirring ethanol by 48 h at 0°C. In the case of the reaction of *cis*-aziridine **24b**-(*c*), two compounds in a 70:30 mixture were observed in the ^{1}H NMR spectra. The CH_3 groups of the two compounds were in 1.35 and 1.44 ppm, respectively. After comparison with reported data, both compounds were identified as *trans* isomers of thiazolidinethiones [63]. The major heterocycle was the *trans*-thiazolidine-thione **12b**-(*t*) and the minor heterocycle, the *trans*-isothiazolidinethione **25b**-(*t*). The ring opening on C3 and C2 of the aziridinium by the aziridinethiocarbamate anion of the intermediate **III** explains the formation of both heterocycles, **Scheme 12**. This aziridinium opening reaction has been observed elsewere [59, 62].

When *cis*-aziridine **24a**-(*c*) was reacted with CS_2, in the same reaction conditions, *cis*-thiazolidinethione **12a**-(*c*) was stereoselectively obtained instead of the expected *trans*-isomer in agreement with the Kellog's method, **Scheme 13**. In this case, the retention of the C1 configuration is explained by attack of the aziridinium thiocarbamate zwitterion **VI** on the benzylic carbon, followed by the closure of the intermediate **VII** to recover the initial C1 configuration.

Crystals of *Cis*-thiazolidinethione **12a**-(*c*) were separated from ethanol and its structure studied by X-ray diffraction analysis, **Figure 14**. The N3—C2(S2)—S1 conjugated system is proposed since the distances are of an intermediate value between a single (1.469 Å) and a double (1.279 Å) N—C bond (N3—C2 = 1.35 Å) and a single (1.789 Å) and a double (1.600 Å) C—S bond (S1—C2 = 1.741Å and S2—C2 = 1.659) [51]. Conjugation makes N3 to be in a sp^2 hybridation, as the angles C(4)—N(3)—C(12) = 119.9(3), C(2)—N(3)—C(4) = 116.2(3), and C(2)—N(3)—C(12) = 121.6(3) show values close to 120°. On the other hand, the five membered ring is almost planar since

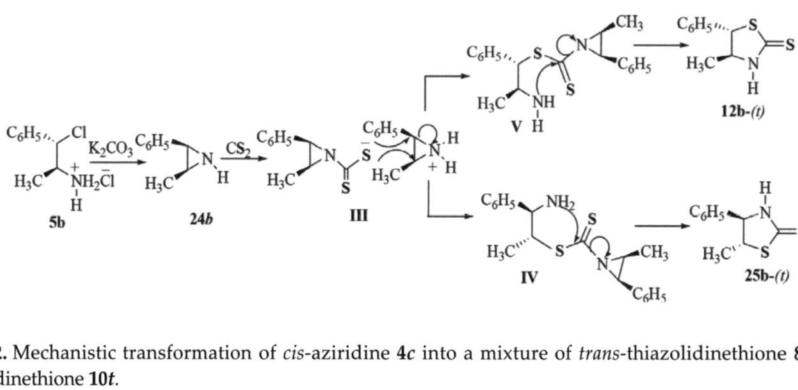

Scheme 12. Mechanistic transformation of *cis*-aziridine **4*c*** into a mixture of *trans*-thiazolidinethione **8*t*** and *trans*-isothiazolidinethione **10*t***.

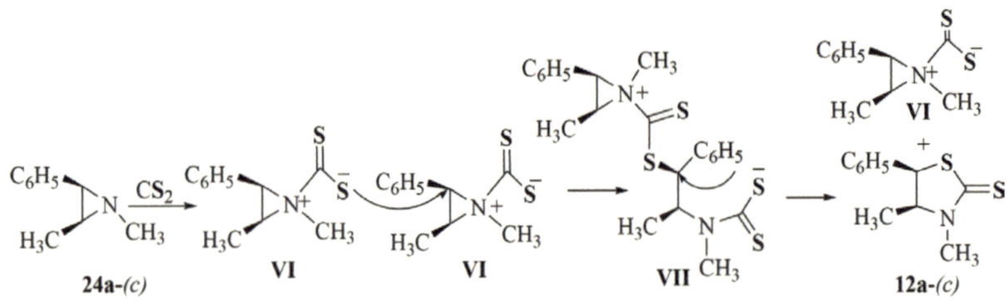

Scheme 13. Mechanistic transformation of *cis*-aziridine **24a**-(*c*) into *cis*-thiazolidinethione **12a**-(*c*).

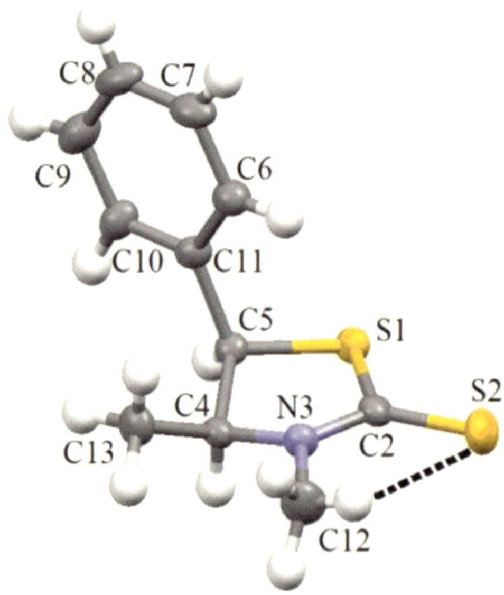

Figure 14. X-ray diffraction structure of *cis*-tiazolidinethione **12a**-(*c*).

the torsion angles N(3)—C(2)—S(1)—C(5) of 5.3(3)°, S(2)—C(2)—N(3)—C(12) of −1.8(5)° are very close to 0°, and S(2)—C(2)—S(1)—C(5) of −177.4(2)°, S(1)—C(2)—N(3)—C(12) of 175.1(3) are close to 180°. An intramolecular contact between a hydrogen atom of the N—CH_3 group and the sulfur atom of the thiocarbonyl group occurred to form a five member ring. The C12H12···S2 distance of 2.72(4) Å [angle of 111(3)°] is in the range for a strong interaction [50].

9. Thiazaborolidines from thioephedrines

The synthesis of N-alkyloxazaborolidines **26–28** derived from ephedrines has been reported (**Figure 15**) [64]. In 1995, we reported the analogous compounds made from thioephedrines, and herein, we report several borohydrides derived from thioephedrine (compounds **12a, 29–37**) following the syntheses depicted in **Scheme 14**.

Hydrolysis of thiosulfate **29** obtained with retention of C1configuration from the substitution reaction of chloride **5** give the disulfide **30, Scheme 14**. The disulfide **30** reacts with BH_3-THF

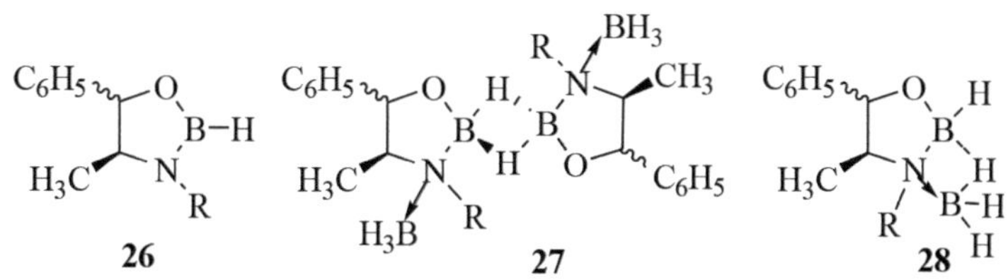

Figure 15. Borolidines from ephedrines.

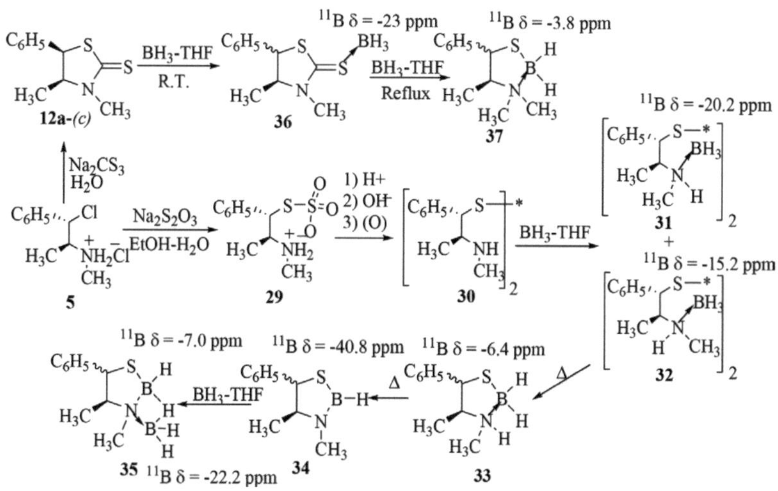

Scheme 14. Thiazaborolidines from ephedrines.

to give a mixture of the stable N-epimers disulfide amine boranes **31** and **32** detected by ^{11}B NMR. This N-epimers are comparable with the N-BH_3 adducts of *pseudo*-ephedrines.

Heating the N-epimers mixture of **31** + **32**, affords the borinic ester **33** as the only product, which in the ^{11}B NMR spectra appears as a triplet [($\delta = -6.4$ ppm, J(BH) = 103 Hz, in $CDCl_3$ or $\delta = -4.5$ ppm, in *THF-d*$_8$]. Two methyl groups in *trans* position was found for borinic ester **33** on the ^{1}H and ^{13}C NMR data. This allows us to assign the configuration at the nitrogen atom. No borinic esters with a BH_2 group derived from ethanolamines as stable compounds have been observed. Borinic ester **33** was distilled *in vacuo*, and on the ^{11}B NMR spectra of the distillated, a mixture of thiazaborolidine **33**, 10%, and thiazaboroline **34** were observed. Slowly elimination of H_2 transforms borinic ester **33** into thiazaboroline **34**. In the ^{11}B NMR spectrum, compound **34** shows a doublet ($\delta = +40.8$ ppm, J(BH) = 154 Hz). From the distilled, a crystal of compound **33** was separated and its X-ray diffraction structure obtained (**Figure 16**). The thiazaboroline **34** reacted with BH_3-THF to afford the N—BH_3 adduct **35** (**Scheme 1**). The structure has been deduced from the ^{11}B NMR data, which indicated a N—BH_3 bond (quadruplet at $\delta = -22.0$ ppm, J(BH)= 71 Hz) and a doublet which is strongly shifted to lower frequencies ($\delta = -7.0$, J(BH)= 148 Hz). A diborane group in which a hydrogen atom from the N—BH_3 adduct is bridging the boron atom of the heterocycle was found. These findings are similar to that found in the *pseudo*ephedrine oxazaborolidine [64a].

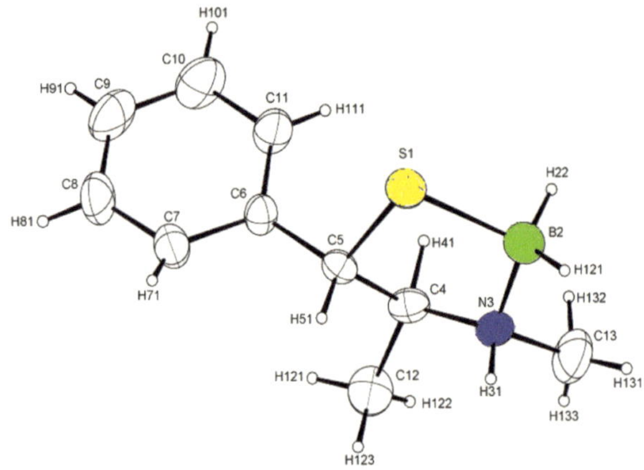

Figure 16. X-ray diffraction structure of borolidine **33**.

Thiazolidine-2-thione **12a**-(*c*) obtained from compound **5a** with sodium trithiocarbonate has been isolated and its reactivity towards BH_3-THF studied, **Scheme 14**. The reaction was followed by ^{11}B NMR, and a S—BH_3 adduct **36** (δ = −23 ppm, broad signal) was first detected. The analysis of ^{1}H NMR spectrum of compound **36** indicates that BH_3 is linked to the thione sulfur atom. Heating the S—BH_3 adduct **36** afforded the thiazaborolidine **37** which is a triplet at δ = −3.8 ppm (J(BH) = I I 1.5 Hz) in ^{11}B NMR spectra. This compound **37** is obtained pure when **12a**-(*c*) is reacted with 3 equivalents of BH_3-THF. The ^{1}H and ^{13}C NMR were recorded. The diasterotopic N-methyl groups show that the nitrogen has a stable configuration, the assignment of the ^{1}H and ^{13}C signals was done by comparison with similar compounds [64b, 65].

Compound **33** has the C-5 atom out of the plane of an envelope conformation of the five member ring. The N—B bond distance is 1.58(1) Å and B—S of 1.922(9) Å. Boron and nitrogen atoms are tetrahedral. The nitrogen atom was found to be of "*S*" configuration, as deduced from the ^{1}H and ^{13}C NMR data. The methyl groups are *trans* position. The angles on the nitrogen atom are close to a sp^3 hybridation, C4—N3—C13 112.3(5)°, B2—N3—C13 111.3(5)°, and C4—N3—B2 112.0(5)°.

10. Biological properties of ephedrines and their derivatives

10.1. Ephedrines

Mao (Ephedra sinica Stapf), which provides similar effects to ephedrine [66, 67], is used as a component of several herbal medicines. It has been utilized in the treatment of cold and allergy [68, 69]. Clinically, it is utilized to lower fever, relieve pain and headaches, control body weight, relieve inflammatory responses [70, 71], and also rheumatoid arthritis [72].

Ephedrine **1a** and *pseudo*ephedrine **2a** are also used to treat cancerous diseases in modern clinical practice, they combined with other preparations relieve arterial spasms, neurotoxic reactions after radiation therapy and chemotherapy [73, 74].

http://dx.doi.org/10.5772/67387

Since 1938, ephedrine was regulated as a drug; however, the herbal source was regulated as a food. However, after the Dietary Supplement Health and Education Act of 1994 (DSHEA) [75] was passed, the herbal products escaped drug regulation. As a consequence, ephedra extracts remains available as a "dietary supplement." However, after years of battling, stimulant combination products (e.g., ephedra and caffeine) are yet available.

10.2. Chloroephedrine derivatives

Previously was demonstrated that N-β-chloroalkylamine derivatives **5a,b** (**Figure 17**) possess a different spectrum of anticancer activities [76]. On the other hand, cytotoxic and antitumor activities of ephedrine and N-β-chloroalkylamine derivatives **5a(1–3)**, **5b(1–3)** were determined [77].

It was found compounds **5b1**, **5b2**, and **5b3** were active in the cytotoxicity test for ^{3}H-thymidine incorporation. The concentrations causing 50% cytotoxicity were in the range 11.0–45.0 mg/mL. However, derivatives **5a1**, **5a2**, and **5a3** were more active.

Compounds **5a2**, **5b2**, **5a3**, and **5b3** investigated *in vitro* suppressed growth of EAC and S-180 tumor strains to various degrees.

It has been shown that the introduction of phosphorus-and sulfur-containing fragments considerably lowers the toxicity of the alkaloids.

With respect to substances **5a1** and **5b1**, they had similar toxicity to *l*-ephedrine, while the other substances were less toxic. It has been found that the replacement of the oxygen atom in the structure of a thio salt of/l-ephedrine **5a3** by a second sulfur atom led to a slight rise in toxicity.

10.3. Dithiocarbamate derivatives

N-methyl-*l*-ephedrinedithiocarbamates derived from *l*-ephedrine **1a** and *d-pseudo*ephedrine **2a** has been obtained. Dithiocarbamic acid derivatives exhibit a broad-spectrum physiological activity [78]. They were found to act as fungicides, herbicides, insecticides, acaricides, zoocides, nematocides, growth regulators, bactericides, etc. Such dithiocarbamic acid derivatives as Carbathion, Cineb, Vegadex, and Cyram have found practical application in agriculture as pesticides.

10.4. Oxazolidines derivatives

Due to the reversibility of the reaction of ephedrines **1–4** with aldehydes or ketones to get oxazolidine heterocycles (**Scheme 15**), these compounds could be used as prodrugs [79]. Some

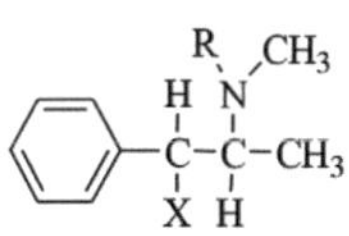

1a X=OH, R=H **2a**
5a1 X=Cl, R=CH_2CH_3 **5b1**
5a2 X=Cl, R=CH_2CH_2Cl **5b2**
5a3 X=Cl, R=NO **5b3**

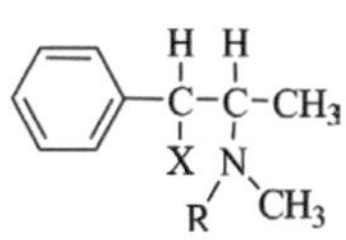

Figure 17. Structures of N-β-chloroalkylamine derivatives **5a,b**.

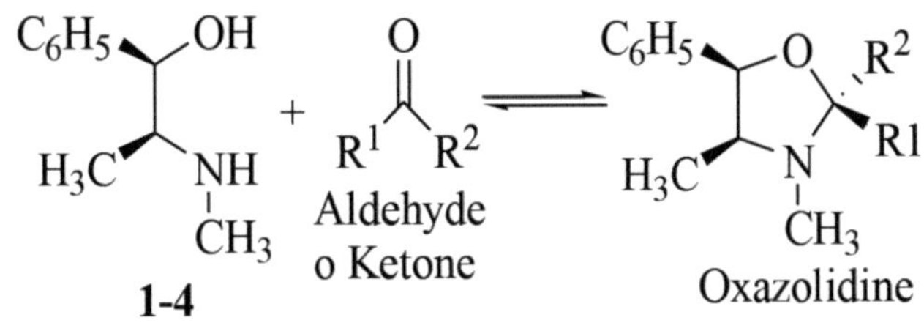

Scheme 15. Oxazolidines from ephedrine.

of these compounds significantly increased locomotor activity in rats at 50-mg/kg dose. The formaldehyde derivative had similar activity as ephedrine. All other compounds were less active.

Four such compounds were tested in rats for ephedrine-like activity using the hyperthermia and anorexia models. The results showed that all of the compounds decreased food intake significantly, but only the acetone and the salicylaldehyde derivatives caused a significant elevation of body temperature [80].

On this bases, we probed the antioxidant and antimicrobial activity of some heterocyclic compounds derived from ephedrines **1–4** previously synthetized.

11. Determination of biological activity of ephedracycles

Ephedrine is a very good pharmaceutical but it acts as central nervous system stimulant, and ephedrine and their derivatives have been used as drugs of abuse so its prescription has been restricted, we proposed ephedra heterocycles as new derived compounds as pharmaceutical candidates with low central nervous system. We decided to prove the antioxidant and antibiotic activities of several heterocycles synthetized in our laboratory represented in **Tables 6** and **7**, respectively.

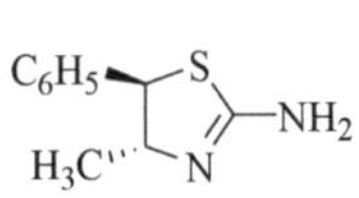

21b(NA)

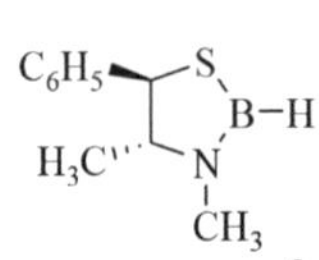

34 ($3.2x10^{-3}$)

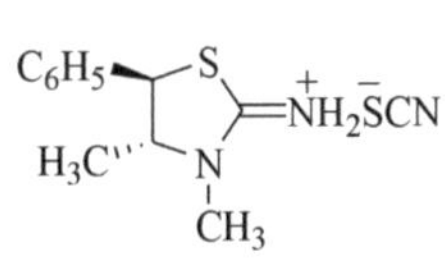

9t (NA)

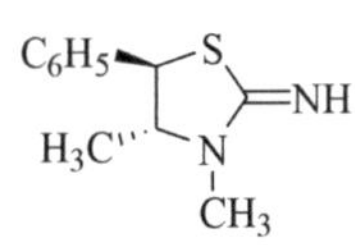

21a (NA)

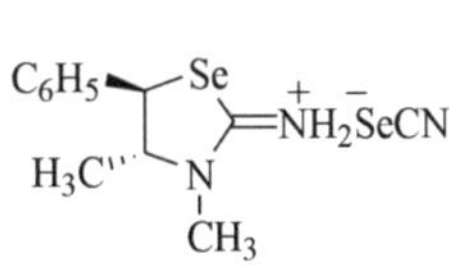

19t ($5.5x10^{-3}$)

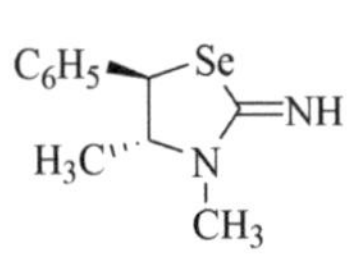

22a (NA)

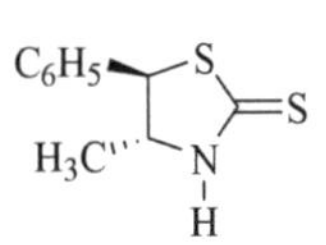

12t ($8.2x10^{-3}$)

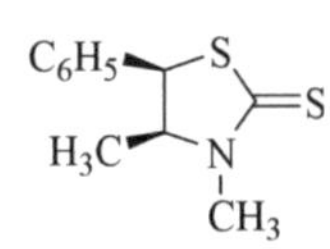

12ac (NA)

Table 6. Antioxidant activity (IC_{50} mol/L) of some synthetized heterocyclic compounds.

Compuesto	*Salmonella typhi*	*Staphylococcus aureus*	*Escherichia coli*	*Bacillus subtilis*
21b	NA	NA	NA	NA
34	**20**	**8**	**10**	**20**
9t	NA	NA	NA	NA
21a	NA	**6**	NA	**9**
19t	NA	NA	NA	NA
22a	NA	NA	NA	NA
12t	NA	NA	NA	NA
12ac	NA	NA	NA	NA

Inhibition in mm.
No actividad found (NA).

Table 7. Antibiotic activity of some synthetized heterocyclic compounds derived from ephedrine **1–8**.

11.1. Antioxidant activity

The DPPH radical scavenging activity of plants was estimated according to the method explained by Cheung, with some modifications. Aliquots of 2 ml of 6×10^{-5} M DPPH methanol were mixed with 50 μL of the extracts. The mixtures were vigorously shaken and left to stand for 10 min under subdued light. The absorbance at 540 nm was measured against methanol as a blank. The decolorization was spectrophotometrically measured at 517 nm. The radical scavenging activity (RSA) was calculated using the equation:

$$\%\mathrm{RSA} = 100 \times \left(1 - A_E/\mathrm{AD}\right)$$

A_E is the absorbance of the solution containing antioxidant extract, whereas AD is the absorbance of the DPPH* solution.

Compounds **34**, **19t**, and **12t** showed antioxidant activity, **Table 6**.

11.2. Antimicrobial activity

Disk diffusion assay: extracts were tested for antibiotic activity against *Escherichia coli, Salmonella thypi, Staphylococcus aureus, Bacillus subtilis,* and *Candida albicans*. About 50 μL of extract was solubilized in ETOH and placed on the surface of the inoculated agar and incubated at 30°C, using antibiotic no. 1 medium, and the antibiotic activity was recorded as the diameter of clear zones of inhibited microbial growth around the paper disk.

The antimicrobial activity was determined using strains *S. aureus, B. subtilis, S. thypi,* and *E. coli* using sensidiscs. The antioxidant activity was measured by the radical 2-2-diphenyl-1-picrihydrazil. Compound **34** presented antimicrobial activity against all microorganisms used and the compound **21a** showed activity only against *S. aureus* and *B. subtilis*, **Table 7**.

Author details

Alejandro Cruz[1*], Itzia I. Padilla-Martínez[1] and María Esther Bautista-Ramírez[2]

*Address all correspondence to: alecruz@ipn.mx

1 Department of Basic Sciences, UPIBI, National Polytechnic Institute, Mexico City, Mexico

2 Department of Pharmaceutical Technology, UPIBI, National Polytechnic Institute, Mexico City, Mexico

References

[1] Nagai, N. Ephedra vulgaris Rich. var. helvetica Hook. et Thomps. Pharm. Ztg. 1887, xxxii, 700.

[2] Kanao, W. l-norephedrine (uber die bestandteila der chinesis chen droge "Ma Huang", VII, Mitterl.) Ber. 1930, 63, 95-98. DOI:10.1002/cber.19300630112

[3] Chen, K. K. A comparative study of synthetic and natural ephedrines. J. Pharmacol. 1928, 33, 219-235. DOI:10.1007/BF01776642. ISSN 1521-0103

[4] Korolkovas, A. and Burckhalter, J. H. Essentials of Medicinal Chemistry; John Wiley and Sons, 606 third Ave., New York, NY 10016, 1976, p. 388-389. DOI: http://dx.doi.org/10.1002/jps.2600660657.

[5] Laccourreye, O.; Werner, A.; Giroud, J.-P.; Couloigner, V.; Bonfils, P.; Bondon-Guittone E. " Benefits, limits and danger of ephedrine and pseudoephedrine as nasal decongestants" European Annals of Otorhinolaryngology, Head and Neck diseases 2015, 132, 31-34.

[6] Goodman, L. S., Gilman, A. "The pharmacological basis of therapeutics", 12th ed., The MacGraw-Hill Co. New York, N. Y., 2011, p. 516. ISBN 978-0-07-162442-8

[7] Helmchen, G., Mulzer, Houben Wegl, J., "Methods of organic chemistry", 4th. Edn.; Eds. Helmchen, G., Hoffman, R. W., Mulzer, J., Schaumanbn, E., Thiemme, Stuttgard, New York, 1995, Vol. *E21a*, pp. 1-146.

[8] (a) Hitchcock, S. R., Casper D. M., Vaughn, J. F., Finefield, J. M., Ferrence G. M., Esken, J. M. Inramolecuar chiral relay at stereogenic nitrogen. Synthesis and application of a new chiral auxiliary derived from (1R, 2S) norephedrine and acetone. J. Org. Chem. 2004, 69, 714-718. doi: 10.1021/jo035325n. (b) Parrott II, R. W., Hitchcock, S. R. Diasteroselective cycloadition of chiral 1-acryloil-2-imidazolidinone and o-quinodimethane generated by reduction of 1,2-bis(bromomethyl)benzene with zinc. Tetrahedron: Asymm. 2007, 18, 988-993. doi:10.1016/j.tetasy.2007.04.014. (c) Kise, N., Mimura, R. Intramolecular chiral relay at stereogenic nitrogen: oxazolidine catalysts derived from ephedra alkaloids. Tetrahedron: Asymm. 2007, 18, 377-382. doi:10.1016/j.tetasy.2006.12.029.

[9] (a) Jahyo Kang*, Dong Soo Kim, Joo In Kim "Enantioselective Addition of Diethylzinc to Aldehydes Catalyzed by A Drug-Unrelated Chiral Amino Thiol and the Corresponding Disulfide" Synlett 1994(10): 842-844. DOI: 10.1055/s-1994-23025. (b) Kang, J., Kim, D. S., Kim, J. H. Enantioselective addition of diethylzinc to aldehydes catalyzed by drugunrelated chiral amino thiol and the correponding disulfide. Synlett 1994, 842-844. doi: 10.1055/s-1994-23025. (c) Fitzpatrick, K., Hulst, R., Kellogg, R. M. Thiol and disulfide derivatives of ephedra alkaloids 2: a mechanistic study of their effect on the addition of diethylzinc to benzaldehyde. Tetrahedron: Asymm. 1995, 6, 1861-1864.

[10] (a) Abramson, S., Laspéras, M., Brunel, D. Design of mesoforous aluminosilicates supported (1R, 2S)-(−)-ephedrine: evidence for the main factors influencing catalytic activity in the enantioselective alkylation of benzaldehyde with diethylzinc. Tetrahedron Asymm. 2002, 13, 357-363. (b) McGhee, A. M., Kizirian J.-C., Procter, D. J. "Synthesis and evaluation of a new polymer-supported pseudoephedrine auxiliary for asymmetric alkylations on solid phase" Org. Biomol. Chem. 2007, 5, 1021-1024.

[11] Ladenburg, A., Oelschägel, C. Uber des pseudoephedrine. Ber. 1889, 22, 1823-1827.

[12] Schmith, E., Bümming U. Ephedrin und pseudoephedrin Arch. Pharm. 1909, 247, 141-149.

[13] Chen, K. K.; Kao, C. H. "ephedrine and pseudoephedrine, their isolation, isomerism, properties, derivatives and synthesis" J. Am. Pharm. Assoc. 1912, 15, 625-639 (b) Nagai, N., Kanao, S. J. Pharm. Soc. Japan 1928, 48, 845.

[14] Gadamer, J. Ueber die Isomerie von Ephedrin und Pseudoephedrin. Arch. Pharm. 1908, 246, 566-574.

[15] Mitchell, W. The acetylation of d-pseudoephedrine and l-ephedrine. J. Chem. Soc. 1940, 1153-1155.

[16] (a) Freudenberg, K., Schoeffel, E., Braun, E. Study of the configuration of ephedrine. J. Am. Chem. Soc. 1932, 54, 234-236. (b) Freudenberg, K., Nikolai, F. Configuration of ephedrine. Annalen. 1934, 510, 223-230.

[17] Leithe, W. Die Konfiguration der ephedrin-basen. Ber. 1932, 65, 660-663.

[18] Leithe, W. Die Konfiguration des optisch aktiven α-phenyl-athylamins, sowre des basen vm thypus des laurdanosins und des tetrahydro-berberins. Ber. 1931, 64, 2827-2832.

[19] (a) Zhu, H. L., Ren, L., Pittman, U. Ch. Jr. Matrix model to predict specific optical rotations of acyclic chiral molecules. Tetrahedron 2007, 63, 2292-2314. (b) Cruz, A., Bautista Ramirez, M. E. Ephedrines and their acyclic derivatives. Curr. Org. Synth. 2011, 8, 901-928.

[20] Späth, E., Göhring, R. Synthesis of ephedrine, pseudoephedrine, their optical antipodes and racemic compound. Monatsh. 1920, 41, 319-338.

[21] Emde, H. Über Diastereomerie I. Konfiguration des ephedrins. Helv. Chim. Acta 1929, 12, 365-376.

[22] Fodor, G., Bruckner, V., Kiss, T., Ohegyi G. Use of acyl migration in separating diastereomeric amino alcohols. J. Org. Chem. 1949, 14, 337-345.

[23] Welsh, L. L. H. Mechanism and stereochemical course of acyl migrations in derivatives of ephedrine and pseudoephedrine. J. Am. Chem. Soc. 1949, 71, 3500-3506.

[24] (a) Pullman, B., Coubeils, J.-L., Courriere, P., Gervois, J.-P. Quantum mechanical study of the conformational properties of phenethylamines of biochemical and medicinal interest. J. Med. Chem. 1972, 15, 17-23. (b) Kier L. B. Molecular orbital calculations of the preferred conformations of histeamine and a theory its dual activity. J. Med. Chem. 1968, 11, 441-445.

[25] Portoghese, P. S. Stereochemical studies on medicinal agents. IV. J. Med. Chem. 1967, 10, 1057-1063.

[26] Bergin, R. Refinement of the structure of (−)-ephedrine hydrochloride. Acta Cryst. 1971, B27, 381-386.

[27] Eran, R. A., Freeman, G. R., Bugg, C. E. Conformational and phosphate binding properties of phenylethanolamines. Crystal structure of ephedrine monohydrogen phpsphate monohydrate. J. Am. Chem. Soc. 1973, 95, 7150-7154.

[28] Earn, R. A., Bugg, C. E. The crystal structure of (−)-ephedrine dihydrogen phosphate. Acta Cryst. 1972, B28, 3662-3667.

[29] Mathew, M., Palenik, G. J. The crystal and molecular structures of (+)-pseudoephedrine and (+)-pseudoephedrine hydrochloride. Acta Cryst. 1977, B33, 1016-1022.

[30] Bailey, N. A., Harrison, P. M., Mason, R. The structure of a bis-(+)-pseudoephedrine complex of copper (II). Chem. Commun. 1968, 10, 559-560.

[31] Carlstrom, D., Bergin, R. The structure of the catecholamines. I. The crystal structure of noradrenaline hydrochloride. Acta Cryst. 1967, B23, 313-319.

[32] Bergin, R., Carlstrom, D. The structure of the catecholamines. II. The crystal structure of dopamine hydrochloride. Acta Cryst. 1968, B24, 1506-1510.

[33] Malone, J. F., Parvez, M. The crystal structures of the ephedrine free bases. Acta Cryst. A 1978, 34, S76.

[34] Close, W. J. The conformations of the ephedrines. J. Org. Chem. 1950, 15, 1131-1134.

[35] (a) Sheldon, Roger, A. "Chirotechnology: industrial synthesis of optically active compounds", Marcel Dekker. New York, 1993, pp. 15-23, 135-137, 253-255. (b) Cain, C. K., "Annual reports in medicinal chemistry", Academic Press Inc., New York, N. Y., 1966, p. 54.

[36] John S. Glasby "Encyclopedia of the alkaloids" Plenium Press, New York and London, 1975, vol.1, p. 518, 1013.

[37] Goodman, L. S.; Gilman, A., "The Pharmacological Basis of Therapeutics", 3rd ed, the Macmillan Co.; New York, N. Y., 1965, p. 516.

[38] (a) Brandt, S. D., Baumann, M. H., Partilla, J. S., Kavanagh, P. V., Power, J. D., Talbot, B., Twamley, B., Mahony, O., O'Brien, J., Elliott, S. P., Archer, R. P., Patrick, J., Singh, K., Dempstera N. M., Cosbey, S. H. Characterization of a novel and potentially lethal

designer drug (±)-cis-para-methyl 4-methylaminorex (4,4′-DMAR, or 'Serotoni') Drug Test. Anal. 2014, 6, 684-695. DOI: 10.1002/dta.1668. (b) McLaughlin, G., Morris, N., Kavanagh, P. V., Power, J. D., Twamley, B., O'Brien, J., Talbot, B., Dowling, G., Mahony, O., Simon D. B., Patrick, J., Archer, R. P., Partillal, J. S., Baumann. M. H. Synthesis, characterization, and monoamine transporter activity of the new psychoactive substance 3′,4′-methylenedioxy-4-methylaminorex (MDMAR). Drug Test. Anal. 2015, 7, 555-564. DOI: 10.1002/dta.1732.

[39] Pfanz, H., Kirchner, G. Ann. 1958, 614, 149.

[40] (a) Trepanier, D. L., Sprancmanis, V., Tharpe, D. S., Krieger, P. E. 5,6-dihydro-4H,1,3,4-oxadiazines. IV. Further studies J. Heterocyclic Chem. 1965, 2, 403-409. (b) Trepanier, D. L., Kriger, P. E., Eble, J. N. Substituted 5,6-dihydro-2-(2-, 3-, and 4-pyridyl)-4H-1,3,4-oxadiazines J. Med. Chem. 1965, 8, 802-807. (c) Trepanier, D. L., Sprancmanis, V., Eble, J. N. 5,6-Dihydro-4H-1,3,4-oxadiazines. V. Base-catalyzed cyclodehydrohalogenation of 2-(β-chloroalkyl)carboxylic acid hydrazides. J. Med. Chem. 1966, 9, 753.

[41] Kalm, M. J. U. S. Patent, 1966, 3, 251, 838.

[42] Kojima, M. Studies in Stereochemistry. XVII. Stereochemical correlation between epimeric dl-2-aminociclohexanol and dl-2-aminociclohexanothiol: participation of neighboring groups in their 1,2-cyclization reactions. Yakugaku Zasshi 1959, 79, 1-10.

[43] Roszkowski, A. P. and Koelle G. B. Enhancement of inhibitory and excitatory effects of catecholamines. J. Pharmacol. Exp. Ther. 1960, 128, 227-232.

[44] (a) Trepanier, D. L., Reifschneider, W., Shumaker, W., Tharpe, D. S. 5,6-Dihydro-4H-1,3,4-thiadiazines[1] J. Org. Chem. 1965, 30, 2228-2234. (b) Trepanier, D. L. U. S. Patent, 1966, 3, 290, 303.

[45] Trepanier, D. L., Wagner, E. R., Harris, G., Rudzik, A. D. 1,4,5,6-Tetrahydro-as-triazines. I. Sulfuric acid catalyzed condenzation of nitriles and hydrazino alcohos· J. Med. Chem. 1966, 9, 881-885.

[46] Murakami, M., Fukumoto, T. Nippon Kagaku Zasshi 1955, 76, 270.

[47] (a) Buur, A., Bundgaar, H. Arch. Pharm., Chem. Sci. Ed. 1987, 15, 76. (b) Yuong-Harvey, J. A. Rae, I. D., Pitman, I. H. Oxazolidines: prodrugs for the delivery of β-aminoalcohols through human skin from aqueous solution. Int. J. Pharm. 1986, 30, 151-160. (c) Chalina, E., Dantchev, D., Georgiev, A., Mitova, K. Ether derivatives of 3-amino-1,2-propanediols, V. Syntheses and pharmacological activities of 5-(cycloalkoxymethyl) oxazolidines and N-substituted derivatives Arch. Pharm. 1986, 319, 598-603.

[48] (a) Schmidt, E. Ueber das Ephedrin und Pseudoephedrin Arch. Pharm. 1914, 252, 89-138. (b) Mitchell, W. The acetylation of d-ephedrine and l-ephedrine. J. Chem. Soc. 1940, 1153-1155.

[49] (a) Dieter R. K.. Deo N., Lagu B., Dieter J. W. Stereo- and regioselective synthesis of chiral diamines and triamines from pseudoephedrine and ephedrine. J. Org. Chem.

1992, 57, 1663-1671. (b) Galindo, A., Orea, L. F., Gnecco, D., Enriquez, R. G., Toscano, R. A., Reynolds, W. F. Synthesis of (2R, 3S)-(−)-2-phenyl-3-methylaziridine. Tetrahedron: Asymm. 1997, 8, 2877-2879.

[50] (a) Cruz, A., Padilla-Martínez, I.I., García-Báez, E.V. Efficient synthesis of cis-thiazolidinethiones derived from ephedrines. Tetrahedron: Asymm. 2011, 22, 394-398. DOI:10.1016/j.tetasy.2011.02.016. (b) Huheey, J. E.; Keiter, E. A and Keiter, R. L. "Inorganic Chemistry" 1978, 4th. Ed.; Eds. Haper Collins College Publishers, Inc., 10 East 53 rd Street, Newyork, N. Y. 10022, p. 300.

[51] Allen, F. H., Kennard, O., Watson, D. G., Brammer, L., Orpen, A. G., Taylor, R. J. Chem. Soc. Perkin Trans. II 1987, S1–S19.

[52] Cruz, A., Geníz, E., Contreras, R. 1,3-heterazolidin-2-one as starting materials for optically active 1,3,2-diazaboroline derived from ephedrines. Tetrahedron Asymm. 1998, 9, 3991-3996.

[53] (a) Gabriel, S., Colman, J. Uber α-phenyl-β-chlorathylanin, $C_6H_5.CH(NH_2).CH_2Cl$. Chem. Ber. 1914, 47, 1866-1873. (b). Clapp, L. B., Watjen J. W. Reactions of ethylenimines. IV. With carbon disulfide. J. Am. Chem. Soc. 1953, 75, 1490-1492. (c) Winternitz, F.; Mouserron, M. and Dennilauler, R. Bull. Soc. Chim. Fr. 1956, 382, 1228-1233. (d) Earley, J. E., Rourke, O. E., Clapp, L. B., Edwards, J. O., Lawes, B. C. Reactions of ethylenimines. IX. The mechanisms of ring openings of ethylenimines in acidic aqueous solutions. J. Am. Chem. Soc. 1958, 80, 3458-3462. (e) Dewey, Ch. S., Bafford, R. A. The reactions of β-aminoalkyl hydrogen sulfates. II. The reaction of sodium β-aminoalkyl sulfate with potassium ethyl xanthate. J. Org. Chem. 1965, 30, 495-500. (f) Kotera, K., Miyazak, S., Takahashi, H., Okada, T., Kitahonoki, K. Aziridine formation by lithium aluminium hydride reduction of oximes. Tetrahedron 1968, 24, 3681-3696.

[54] (a) Evans, D. A., Mathre, D. J., Scott, W. L. Asymmetric synthesis of the enkephalinase inhibitor thiorphan J. Org. Chem. 1985, 50, 1830-1835. (b) Fodor, G., Stepanovsky, J., Kurtev, B. Optisches Drehungsvermögen und Konformation, 3. Mitt.: über die Verhältnisse Beicis-Ringschlubreaktionen epimerer Aminoalkohole der Ephedrinreihe Monatsh. Chem. 1967, 98, 1027-1042.

[55] Cruz, A., Macías-Mendoza, D., Barragán-Rodríguez, E., Tlahuext, H., Nöth, H., Contreras, R. Synthesis of optically active boroxazolidine, borathiazolidine and their N-borane adducts from the corresponding 2-imino-heterazolidines. Tetrahedron Asymm. 1997, 8, 3903-3911.

[56] Jin, M.-J., Kim, Y.-M., Lee, K.-S. Enantioselective addition of diethylzinc to aldehydes in the presence of chiral aprotic ligands. Tetrahedron Lett. 2005, 46, 2695-2696.

[57] Flores-Parra, A., Suárez-Moreno, P., Sánchez-Ruíz, S, Tlahuextl, M, Jaen-Gaspar, J., Tlahuext, H., Salas-Coronado, R, Cruz, A., Nöth, H. Contreras, R. Clorination reactions of ephedrines revisited. Stereochemistry and functional group effect on the reaction mechanisms. Terahedron Asymm. 1998, 9, 1661-1671.

[58] Cruz, A., Contreras R., Padilla-Martínez, I. I. and Juárez-Juárez, M. 1,3-heterazolidines-2-heteroinsaturated compounds derived from ephedrines. Tetrahedron Asumm. 2006, 17, 1499-1505. DOI:10.1016/j.tetasy.2006.05.009.

[59] Cruz, A., Padilla-Martínez, I. I., García-Báez, E. V., Contreras, R. Reactivity of chlorodeoxypseudoephedrines with oxo-, thio-, and selenocyanates. Tetrahedron: Asymm. 2007, 18, 123-130. DOI:10.1016/j.tetasy.2006.12.012.

[60] Cruz, A., Flores-Parra, A., Tlahuext, H., Contreras, R. Thiazaborolidines adducts derived from thioephedrines Tetrahedron: Asymm. 1995, 6, 1933-1940.

[61] Brunet, E., Carreño, M. C., García-Ruano, J. L. Stereospecific synthesis of thiazolidine-2-thiones. Heteocycles 1985, 23, 1181-1195.

[62] Cruz, A., Padilla-Martínez, I. I., García-Báez, E. V. An efficient synthesis of aziridines from ephedrines Tetrahedron: Asymm. 2010, 21, 909-913. DOI:10.1016/j.tetasy.2010.05.007.

[63] Delaunay, D., Toupet, L., Corre, M. L. Reactivity of beta-amino alcohols with carbon disulfide study on the synthesis of 2-oxazolidinethiones and 2-thiazolidinethiones. J. Org. Chem. 1995, 60, 6604-6607.

[64] (a) Tlahuext, H., Contreras, R. Diborane derivatives from N-alkyloxazaborolidines. Tetrahedron Asymm. 1992, 3, 1145-1148. (b) Tlahuext, H., Sanliesteban, F., Baez, E. G., Contreras, R. Determination of the nitrogen configuration in ephedrine N-boranes and a study of their assisted stereospecific deuteration of NH Tetrahedron Asymm. 1994, 5, 1579-1588.

[65] (a) Contreras, R., Santiesteban, F., Paz-Sandoval, M. A., Wrackmeyer, B. N-borane adducts of oxazolidines derived from ephedrine and pseudoephedrine. Study of stereochemistry by nuclear magnetic resonance. Tetrahedron 1984, 40, 3829-3838. (b) Paz-Sandoval, M.A., Santiesteban, F., Contreras, R. Carbon-13 NMR studies of oxazolidines, N-borane-oxazolidines and oxazolidinium salts derived from pseudoephedrine: steric effect of the BH_3 and CH_3 groups on the chemical shifts. Mag. Reson. Chem. 1985, 23, 428-431.

[66] Bielory, L. Complementary and alternative interventions in asthma, allergy, and immunology. Ann Allergy, Asthma Immunol. 2004, 93, S45–S54.

[67] Gurley, B. J., Gardner, S. F., White, L. M., et al. Ephedrine pharmacokinetics after the ingestion of nutritional supplements containing ephedra sinica (ma huang). Therap. Drug Monit. 1998, 20(4), 439-445.

[68] White, L. M., Gardner, S. F., Gurley, B. J., et al. Pharmacokinetics and cardiovascular effects of ma-huang (ephedra sinica) in normotensive adults. J. Clin. Pharmacol. 1997, 37, 116-122.

[69] Saito, S., Maruyama, Y., Kamiyama, S., Nakahata, N., Ohizumi, Y. Ephedrae herba in mao-bushi-saishin-to inhibits IgE mediated histamine release and increases cAMP content in RBL-2H3 cells. J. Pharmacol. Sci. 2004, 95, 41-46.

[70] Shekelle, P. G., Hardy, M. L., Morton, S. C., Maglione, M., Mojica, W. A., Suttorp, M. J., et al. Efficacy and safety of ephedra and ephedrine for weight loss and athletic performance: a meta-analysis. J. Am. Med. Assoc. 2003, 289, 1537-1545.

[71] Hikino, H., Konno, C., Takata, H., Tamada, M. Antiinflammatory principle of ephedra herbs. Chem. Pharma Bull.1980, 28, 2900-2904.

[72] Yeom, M.J., Lee, H.C., Kim, G.H., Lee, H.J., Shim, I., Oh, S.K., et al. Antiarthritic effects of ephedra sinica Stapf herb-acupuncture: inhibition of lipopolysaccharide-induced inflammation and adjuvant-induced polyarthritis. J. Pharmacol. Sci.2006, 100, 41-50.

[73] Boada, S.; Solsona, B.; Papaceit, J.; Saludes, J. and Rull, M. "Hypotension refractory to ephedrine after sympathetic blockade in a patient on long-term therapy with with tricyclic antidepressants." Rev. Esp. Anestesiol. Reanim. 1999, 46, 364-366.

[74] Punzengruber, C., Schwart, H. P., Weissel, M. Acta Med. Austriaca 1986, 79-81.

[75] U.S. Food and Drug Administration, Center for Food Safety and Applied Nutrition, Dietary Supplement Health and Education Act of 1994.

[76] Leont′ev, V. B.; Yusupov, N. U.; Mukhamedkhanova, S. I.; Kuznetsova, N. N.; Konovalova, N. P.; Vasiléva, L. S.; and Aslanov, Kh. A. Khimioter. Opukholei′v SSSR 1980, 58.

[77] Kuznetsova, N. N., Khashimova, Z. S., Sadikov, A. A. Effect of ephedrine and its derivatives in KML cell culture and on tumor strains of animals. Chem. Nat. Componds. 2004, 40, 258-260.

[78] Halasi, T., Nesterova, M. V., Severin, E. S., Italasi, P. Study of the influence of fungicidal substituted trithiocarbamates on the system of second messengers and the proliferative activity of mammalian cells. Pharm. Chem. J. 1993, 27, 731-733.

[79] Young-Harvey, J. A., Rae, I. D., Pitman, I. H. Oxazolidines prodrugs for the delivery of β-aminoalcohols trough human skin from aqueous solution. Int. J. Pharm. 1986, 30, 151-160. DOI: 10.1016/0378-5173(86)90075-X.

[80] Walker, R. B., Wood, N. D., Mateen, A. M., Sharks, E. Efects of oxazolidines derived from (−)-ephedrine in the rat. Gen. Pharmacol. Vasc. Syst. 1992, 23, 729-732.

Chapter 3

The Local Anesthetic and Pain Relief Activity of Alkaloids

Havva Sayhan, Serbülent Gökhan Beyaz and
Ahmet Çeliktaş

Additional information is available at the end of the chapter

http://dx.doi.org/10.5772/intechopen.69847

Abstract

Alkaloids have been known for several centuries and have been mainly obtained from natural sources that presented important properties with biochemical, pharmacological, and medical effects in living organisms. Alkaloids are derived from amino acids like other important molecules in the functioning of life in our body. Hence, alkaloids are considered as pharmacologically important. Alkaloids are secondary metabolites widely distributed in leaves, stem, root, and fruits of plants which synthesize them. However, administration and consumption of them at right doses are beneficial in terms of health; excess doses will be definitely poisonous and may cause even death. The pharmacological activities of alkaloids are quite diverse. They are important natural products with a wide range of medicinal properties including relief of pain (e.g., morphine), analgesic (e.g., codeine), antiarrhythmic (e.g., quinidine), antibacterial (e.g., chelerythrine), antiasthma (e.g., ephedrine), cholinomimetic (e.g., galantamine), and vasodilatory (e.g., vincamine).

Keywords: alkaloids, local anesthetics, pain relief, analgesia, natural products, mechanism of action, bioorganic chemistry

1. Introduction

MERGEFORMAT plants have been used from ancient times as an excellent source of pharmaceutical compounds in the treatment of diseases and have solid impact on human health. Modern chemistry and material sciences have identified the compounds with pharmacological properties for the development of new drugs. Hence, they have played a major role in drug discovery as natural products and synthetic materials which do not exist in the nature and can only be produced.

The presence of some chemical substances in plant tissues produces a physiological action on the body and regulates the metabolic activity. For example, these chemicals give plants their color or act as defense system based on their toxic effects. Alkaloids have been known for several centuries and have been mainly obtained from natural sources that presented important properties with biochemical, pharmacological, and medical effects in living organisms. Alkaloids are secondary metabolites widely distributed in leaves, stem, root, and fruits of plants which synthesize them. However, administration and consumption of them at right doses are beneficial in terms of health; excess doses will be definitely poisonous and may cause even death. Many alkaloids extracted from plants showed anticholinesterase activity [1, 2] and antioxidant [3], anxiolytic [4], antimicrobial [5], anti-HIV [6], antiparasitic [7], anti-inflammatory [8], and antidepressant [9] properties. After the recognition of vital importance of these nitrogenous secondary metabolites, they are found in animals too [10]. In spite of the investigations on the biological activity of some alkaloids, only a few could be produced commercially due to their complex chemical structures. The synthetic production of some important alkaloid is more common than their isolation from plants. In addition, there are other synthetic compounds that are closely related to the natural alkaloids but lack of some typical properties (e.g., homatropine) [11].

Alkaloids display antimicrobial and antiparasitic properties, act as narcotics, can alter DNA, have an important role in the immune systems, and treat cardiovascular and metabolic disorders, inflammation, infectious diseases, and miscellaneous problems [10, 12–19]. Therefore, modern medicine tends to produce pharmaceuticals on the basis of natural alkaloids and alkaloids with modified structure such as aconitine, ajmaline, berberine, morphine, caffeine, theophylline, ephedrine, atropine, scopolamine, reserpine, and pilocarpine. There are several morphine-containing drugs that are used in cases of surgical operations and postoperative treatments (i.e., Morphalgin, Spasmofen, and Morphin) [20, 21].

2. History

The human race has used alkaloids in the scope of medicinal and pharmaceutical importance even before their discovery as chemical molecules. There are records from the Babylonian, ancient Hebrew, Egyptian, Chinese, Greek, and Assyrian times, such as Sanskrit writings and the works of Hippocrates, showing that they were all familiar with medicinal plants and alkaloids to improve the health in the early 5000 BC. The most prominent examples of medicinal plants based on the earlier beliefs and knowledge were myrrh, opium, cannabis, aloes, cassia, and hemlock. For example, native people of America and tribes in the Amazon used the alkaloid quinine in herbal before its official use in the cure of uncomplicated malaria in 1638. The latex of opium poppy (*Papaver*) was already in use in the Middle East and Greece at 1200 BC. The roots of the mandrake plant were used because of sedative properties from the time of Hippocrates (ca. 400 BC). All alkaloids cause a physiological effect on the human body, and different poisonous alkaloids such as the ergot alkaloid, aconitine, and tubocurarine have been used for toxic purposes in ancient time. The poisonous alkaloids were prepared from plants (also from animal sources) and used as poisons for arrows by hunters in different parts of the world. Literature also refers that some alkaloids including aconitine, atropine, colchicine, coniine, ephedrine, ergotamine, mescaline, morphine, strychnine, psilocin, and psilocybin have been used in executions. Philosopher Socrates was sentenced to death by consuming a cup of hemlock which contains the alkaloid coniine in 399 BC [11, 22–31].

Despite the fact that alkaloids have been in use for ages, actual studies on isolation from plants as relatively pure compounds occurred only in the beginning of the 1800s. In the late 1700s, Lavoisier and others introduced the new developments in chemistry that leads many chemists to try to isolate the active ingredients in plants. The German pharmacist's apprentice Friedrich Sertürner discovered the morphine in opium poppy by dissolving it in acid and neutralizing it with ammonia in 1804 for the first time in history. Morphine is considered for clinical purposes primarily a safe and effective way of pain relief. Although morphine is used clinically, drug addiction began with morphine and related substances. When Sertürner proved that the substance (morphine) he had isolated was indeed responsible for the actions of opium, chemists had isolated several other medically important substances such as quinine, strychnine, and caffeine around that time. While the first use of the name "alkaloids" was in 1819 by a German chemist Carl F.W. Meissner, the term had an extensive usage after the publication of an article in the chemical dictionary of Albert Ladenburg in the 1880s [11, 22–24, 32–37].

3. Properties

An exact and precise definition is somewhat difficult because of differences between the research fields of biology, medicine, and chemistry. The term alkaloid (alkali + oid) means alkali-like substance, and names of alkaloids end in -ine (e.g., atropine, cocaine, and morphine). Many methods have been proposed for the classification of alkaloids such as pharmacological, taxonomic, chemical, and biosynthetic classification. Although chemical classification based on the carbon skeleton is the most accepted way of classification, here we will use the pharmacological one depending on the physiological response. Alkaloids are a class of organic compounds that is made up of carbon, hydrogen, oxygen, and nitrogen. In addition, sulfur and rarely other elements such as chlorine, bromine, and phosphorus are also included in the alkaloids. Primary nitrogenous metabolites including amino acids and polymer of amino acids serve as precursors for alkaloids and are usually largely retained in the structure where the nitrogen in the alkaloid molecule is derived. In particular, alkaloids are synthesized from amino acids mostly derived from four different amino acids: lysine, phenylalanine, tyrosine, and tryptophan. To this respect alkaloids show similar structural features to the originated amino acid (shown in **Figure 1**). Alkaloids usually have one or more heterocyclic ring structure which includes at least one nitrogen atom in the ring. Nevertheless, the nitrogen atom is not within a carbon ring in some alkaloids such as mescaline. They are alkaline in nature due to the nitrogen content that they can absorb acid or hydrogen ions making them a base; however, some do not exhibit alkaline properties. The fact remains that the precise position of the nitrogen atom in the carbon ring or molecule varies with different alkaloids that determines the properties of them [24, 38–40].

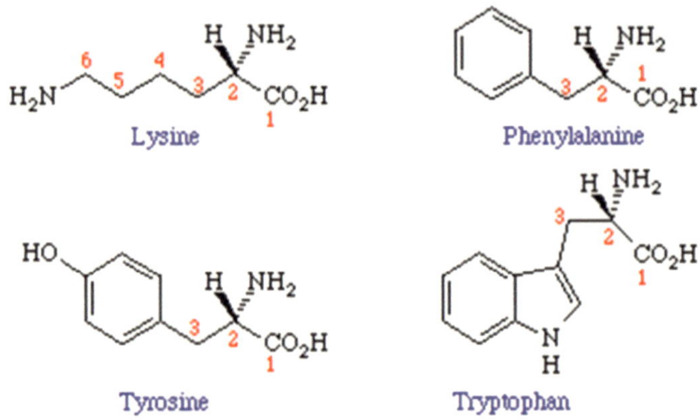

Figure 1. The amino acids used for alkaloid biosynthesis.

From the first impression, the term "secondary metabolite" looks like it is used to describe little importance to plant metabolism or unrelated compounds, although they have broad functional spectrum of specialized metabolism such as defense mechanisms and serve as bioactive compounds for drug discovery, and also other roles have not yet been clearly understood in the plant. It is a matter of controversy among scientists how alkaloids as secondary metabolites make contribution to plants. Some believe that the presence of alkaloids discourages insects and animals from eating plants. Nonetheless, the general properties of alkaloids could be listed as colorless, crystalline solids and nonvolatile, optically active, bitter taste, soluble in organic solvents (i.e., alcohol, ether, chloroform). On the contrary their corresponding salts are highly soluble in water. There are a few liquids such as coniine and nicotine and even a few colored such as berberine is yellow [24, 40, 41].

Alkaloids can form ionic bonds with phenolic hydroxyl groups; hydrogen bonds with hydroxyl groups, carbonyl, or keto groups; and van der Waals and hydrophobic interactions with lipophilic compounds. These interactions will lead a conformational change in the protein structure that is usually associated with a loss or reduction in the bioactivities of proteins (enzymes, receptors, ion channels, hormones, etc.). Apart from other secondary metabolites, alkaloids more specifically interact solely with a single particular target. For example, neurotransmitters and alkaloids both derive from amino acids as a consequence consist of structural similarities and are considered as analogs of each other. Alkaloids can bind to neuroreceptors, either activate (agonists) or inactivate (antagonists) them, and inhibit or activate ion channels such as the Na^+, K^+, and Ca^{2+} channels. Alkaloids have shown significant action for the treatment of such dangerous human diseases as cancer, AIDS, and lung diseases [22, 40, 41]. Neurodegenerative diseases (ND) such as Alzheimer's (AD) disease and dementia primarily affect the neurons in the human (central nerve system) which control many direct body functions and the behavior. The alkaloids show promising pharmacological activities for the treatment of neurodegenerative diseases such as Alzheimer's disease. Clinical studies of galantamine, which is an alkaloid obtained from daffodil (*Narcissus tazetta*), snowdrop (*Galanthus nivalis*), and snowflake (*Leucojum aestivum*), report the capability of stimulating nicotinic receptors that further enhance cognition and memory [42–49].

4. Plants as a source of alkaloids

There is an enormous range of chemical compounds present in plants where alkaloids take place as a content of secondary metabolites. Alkaloid content is inhomogeneous and usually within a few percent over the plant tissues. Further, most plants produce a few types of closely related alkaloids, whereas some of them may contain several alkaloids in different tissues of the same plants. Therefore, firstly there is a need on the extraction of their mixture and then separation of individual alkaloids. Medicinal bioactive compounds from raw extracts of the plants can be isolated by acid-base extraction. Pure alkaloids are not readily soluble in water, but found to be fairly soluble in organic solvents (ether, chloroform, alcohol, and oils), relatively nonpolar solvents (hexane, benzene, petroleum ether). On the contrary, alkaloids are mostly present in the corresponding salt form in the plant and almost freely soluble in water, mostly insoluble or relatively less soluble in many of organic solvents. As a group, alkaloids are easily extracted in acidic forms because of their basic and lipid properties and separated from other water-soluble materials [32, 33, 38, 49–52].

Usually, the dried and powdered plant source is extracted with lipophilic or nonpolar organic solvents such as 1,2-dichloroethane, chloroform, diethyl ether, benzene, or petroleum ether. Since the alkaloids present in the plant sources as the salt of acids, they are exposed to an alkaline medium to convert the alkaloid salts to the corresponding alkaloid bases. The extraction of the alkaloid bases from the bulk of the crude alkaloid solution is achieved again with a nonpolar solvent. The impurities from the plant in solution are dissolved by dilute aqueous acid and are washed away with water. Alkaloids have different solubilities in certain solvents and different reactivities with certain reagents because of the structural diversity [22, 53, 54].

The alkaloid-based drugs used today are of plant origin, and screening of plant extracts for alkaloids and other pharmacologically active compounds is still in progress for new drug discoveries. Chemists extensively investigated production of alkaloids in plants on a large scale, to make many derivatives of these natural compounds and improve technologies correlated to chemical preparation. The discoveries of high-value chemical compounds in plants serve as model structures for synthetic drugs and allow the large-scale production of them with improved properties. Recently, the deep sea bioenvironment is considered an extremely rich source of novel bioactive alkaloids since marine natural products represent a fascinating example of the large variety of secondary metabolites [22, 40, 55].

5. Fundamentals of alkaloid pharmacology

The most useful portion of the plant (e.g., roots and seeds, stem, leaves, bark, milky exudate, etc.) has gathered as drug which is called as crude drugs and has a long-standing place in medicine. In fact, the comprehensive knowledge on the crude drugs is the basis of pharmaceutical sciences. A number of alkaloids are used as drugs owing to the important pharmacological bioactivity in human bodies. In addition to natural alkaloids, synthetic and semisynthetic alkaloids are discovered by chemical synthesis and modifications of natural alkaloids. Thus, synthetic and semisynthetic alkaloids are biologically more active with improved properties. For example, morphine potency has been dramatically increased via the addition of a 14-hydroxy group to the morphine alkaloid structure. Moreover, scientific achievements have allowed discrimination of chemical structural differentiations (angle of valence of C, N, H, and moiety) of the alkaloid molecules that can lead into considerable changes in biological activities. Therefore, the interdisciplinary research of bioorganic chemistry, biology, and pharmacology is becoming more important [41, 56–58].

For many years, scientific research has been tried to understand the nature and the function of alkaloids in plant metabolism. The role of alkaloids can be explained based on the functions of these compounds inside and outside the organism producing them. Alkaloids play a protective or chemical defense role in interaction with other organisms. However, it is not entirely clear if this ecologically important role is a basic function of these compounds. New researches suggested this role may be a secondary function of alkaloids in connection with the regulation of metabolism as the result of gene expression. Moreover, alkaloids have a fundamental role inside the organism in which they occur and produced by the activity of the organism's genes, enzymes, and proteins (i.e., quinolizidine alkaloids). Alkaloids are able to self-regulate by changing their structural chemical configurations and biological activity in different cell conditions according to pH changes.

Alkaloids and alkaloid-containing drugs became critical components of the pharmacology for clinical applications due to the tremendous healing capability [30]. Many methods have been proposed for the classification of alkaloids as pharmacological, taxonomic, chemical, and biosynthetic classification. Although chemical classification based on the carbon skeleton is the most accepted way of classification, here we will use the pharmacological one depending on the physiological response. In medicine, alkaloids mainly exhibit marked pharmacological activity in some serious disorders like cardiovascular and metabolic disorders, cancer, blood pressure, inflammation, infectious diseases, neurodegenerative diseases, and miscellaneous problems. They have been used as bronchodilator, cardiac stimulant (quinidine), muscle relaxant, pain killer, analgesic (codeine), antioxidant, anticancer (berberine), antimicrobial and amebicidal, anti-inflammatory, central nervous system stimulants or depressants, sympathomimetics, purgative, vasodilator (vincamine), etc. Various pharmacological alkaloids are used in medicine with some examples as listed: atropine is widely used as an antidote to cholinesterase inhibitors and also used in drying cough secretions; morphine and codeine are narcotic analgesics, and codeine is also an antitussive agent; colchicine is used as a gout suppressant; caffeine is a central nervous system stimulant further used as a cardiac and respiratory stimulant and as an antidote to barbiturate and morphine poisoning; emetine is used in the treatment of amebic dysentery and other protozoal infections; epinephrine or adrenaline is used as a bronchodilator and cardiac stimulant and antiallergic, anesthesia, and cardiac arrest; vincristine is used as anticancer drug; and ephedrine is used in blood pressure. Galantamine plays a vital role in treating cognitive disorders that cognitive dysfunction is a major health problem in the twenty-first century, by influencing the function of receptors for the major inhibitory neurotransmitters [22, 30, 54, 55, 59, 60].

http://dx.doi.org/10.5772/intechopen.69847

6. Alkaloids with anesthetic effects and the related mechanisms

Local anesthetics are the most effective drugs used for the provision of anesthesia and analgesia both intra- and postoperatively in medicine. They are also used to decrease temperature, touch proprioception, and skeletal muscle tone. Alkaloids in local anesthetics have been used in a variety of clinical situations such as topical application to the skin or mucosa membranes; injectable agents for peripheral, central, or spinal nerve block; and also anorectal or ophthalmic use. Local anesthetics consist of a lipophilic (soluble in lipids) aromatic ring connected to a hydrophilic amine ring (amide group) in the molecular structure. In all clinically used local anesthetics (except cocaine), these lipophilic and hydrophilic groups bind via an amide or ester, and the nature of this bond determines many of the properties of the agent (**Figure 2**). Cocaine, procaine, tetracaine, chloroprocaine, benzocaine, and amethocaine can be given as examples to esters, whereas lidocaine (also known as lignocaine), bupivacaine, mepivacaine, prilocaine, etidocaine, dibucaine, ropivacaine, and levobupivacaine for amides. As can be noticed, the suffix "-caine" is a common ending for drugs containing alkaloids. The ester bond is less stable than amide bond so it can easily be broken in solution and cannot be stored for as long as amides. Amino esters undergo hydrolysis to derivatives of para-aminobenzoic acid (PABA) during the metabolic process in the blood which cause allergic reactions that range from urticaria to anaphylaxis. In contrast, amino amides are metabolized by enzymes in the liver, have long duration of activity, are heat stable, also rarely trigger an allergic reaction, and therefore are more commonly used than esters.

When using the local anesthetics, it should be adjusted according to the duration of the surgical procedure and the anticipated degree of pain. For example, a short-acting agent (e.g., mepivacaine) will be useful for creation of excellent intraoperative conditions with minimum postoperative pain for a short operation. On the other hand, a long-acting anesthetic (e.g., ropivacaine) selection will be appropriate for a rotator cuff repair which involves a greater degree of postoperative pain. A given local anesthetic has different block durations and onset depending on the nerve or plexus blockade.

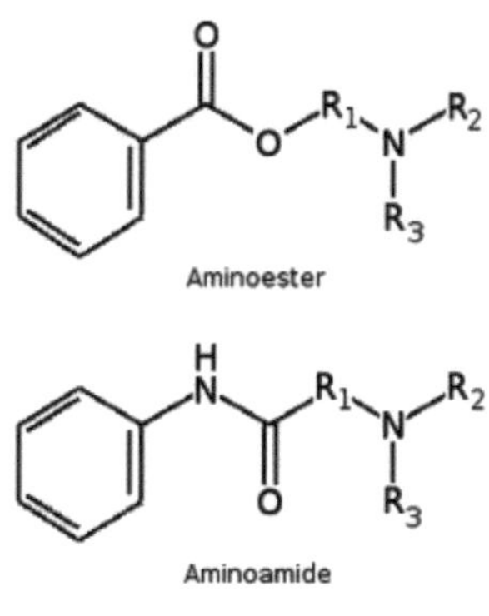

Figure 2. Para-aminobenzoic derivatives of local anesthetics (hydrolyzed by pseudocholinesterase in plasma).

All local anesthetics (except cocaine) are vasodilators where vasodilation occurs via direct relaxation of peripheral arteriolar smooth muscle fibers. The enhanced vasodilator activity of a local anesthetic agent results in shorter duration of action because of faster absorption. A vasoconstrictor counteracts this vasodilatation by providing delayed vascular adsorption of local anesthetics and increases the duration of contact with nerve tissues. Epinephrine vasoconstricts arteries, in concentrations of 5 mcg/ml (1:200,000), and is commonly used to decrease the adsorption of lidocaine when combined with local anesthetics solution. This dosage of epinephrine will significantly reduce nerve blood supply of lidocaine regardless of the site of administration. Decrease in the absorption provides increased neuronal uptake, enhanced quality of anesthesia, and prolonged duration of action [22, 23, 41, 56, 59, 61].

The first local anesthetic used in the clinical practice was cocaine, an alkaloid from the leaves of *Erythroxylum coca*. The natives of South America have chewed the leaves of *E. coca* which grows wild in the Andes Mountains in Peru for the stimulant effect, and the systemic effects of chewing these leaves had reached Europe from the time of the Spanish conquest. In 1860, Niemann isolated the pure alkaloid responsible for the properties of the coca leaves and produced pure white crystals which he named cocaine. He noted that it had a bitter taste and produced numbness of the tongue, rendering an almost devoid sensation. A Russian physician Vassily von Anrep in 1880 demonstrated that subcutaneous injection of cocaine produces sensory block where the skin is insensitive to the prick of a pin. This had attracted the attention of Sigmund Freud, and he conducted a study of the anesthetizing properties of cocaine with assistance of his friend Carl Koller. Koller introduced its topical use into ophthalmology and found that cocaine is able to block signal conduction in nerves which led to its rapid medical use as a "local anesthetic" in both dentistry (1884) and in surgery (1885) in spite of its dangers (dangerous effect on the central nervous system, drug of addiction, mydriasis in eye surgery) [62–64].

In 1885, Leonard Corning produced spinal anesthesia both in a dog and a patient and produced block of the lower half of the body. Even though he suggested its feasibility, it could be performed in surgery several years after that. Quincke showed that lumbar puncture was a practical procedure in 1891 and the first spinal blocks for surgery was performed by August Bier in 1898.

The early use of cocaine was largely limited to topical application because of the toxicity of cocaine, difficulties in sterilization, drug addiction, and short duration of action. Later on, others have been discovered with less abuse potential, low toxicity, and safer drugs like procaine and procainamide. The latest is lidocaine (synthetic local anesthetic considered prototype) where all new local anesthetics like amino amides have been introduced into clinical practice (**Figure 3**) [46, 47, 65].

6.1. Mechanism of action

Local anesthetics block the generation and the conduction in peripheral nerves by disruption of ion channel function within the neuron cell membrane preventing the transmission of the neuronal action. This is achieved by specific binding of alkaloids in anesthetics reversibly to sodium (Na+) channels, preventing the passage of Na+ through Na+ channels, and interaction

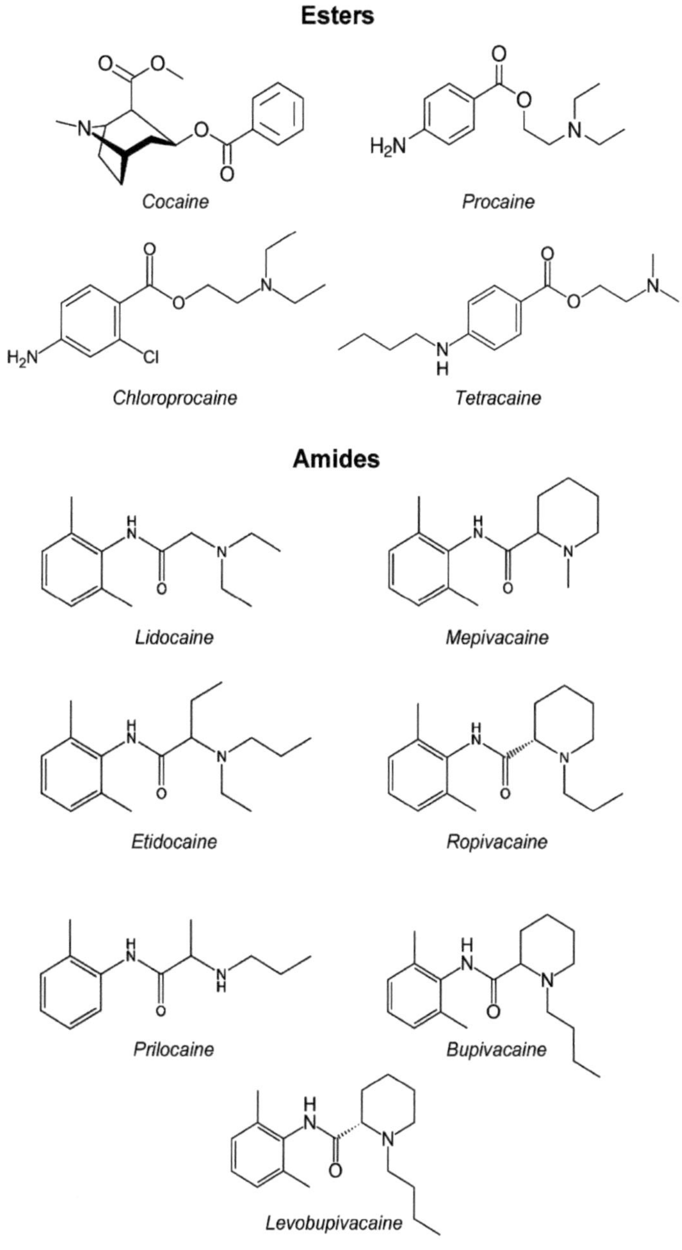

Figure 3. Chemical structures of ester and amide local anesthetics.

with receptors, holding them in an inactive state so that no further depolarization can occur. Local anesthetics primarily have the ability to bind to sodium channel pore when it is in activated state, although may bind during the resting inactivated state to make it impermeable to Na+ [62, 66–68].

Sodium influx through these channels is necessary to function within the neuron cell membrane for the transmission of the neuronal action potential and subsequent propagation of an impulse along the course of the nerve. Thus, the loss of sensation in the area supplied by the nerve provides loss of pain, temperature, touch, proprioception, and then skeletal muscle tone, respectively. When local anesthesia is used, people may still feel touch but not pain.

Physiologic activity of alkaloidal local anesthetics depends on lipid solubility, affinity for protein binding, percent ionization at physiologic pH, and vasodilating properties. Lipid solubility is directly in correlation with the potency of the local anesthetic where the nerve cell membrane is composed of lipid and a depot of drug forms in the perineural lipid-rich tissues.

Through the lipid solubility, lidocaine penetrates in nerve membranes and binds to acute-phase protein of α1-acid glycoprotein (AAG) which is responsible for binding lidocaine and other basic drugs. Although lidocaine can bind to albumin (low affinity, high capacity), AAG (low capacity, high affinity) is the major drug-binding macromolecule even with lower plasma concentration. Albumin is considered as the principal binding protein for acidic compounds, and AAG is the principal binding protein for basic drugs [69–72].

Alkaloids are weak bases and exist in ionized and nonionized forms which are defined by the dissociation constant (pKa) of a weak base and the pH of the tissues (usually at range 7.6–8.9). Alkaloids are able to self-regulate by changing their structural chemical configurations and biological activity in different cell conditions according to pH changes. The neutral base form of the local anesthetic is more lipophilic that makes pKa greater than 7.4 for most local anesthetics. A decrease in pH increases the proportions of ionized form and results the delayed onset of action. In the presence of inflammation, local anesthetics are less effective, and onset of action is slow. On the contrary, an increase in pH results increased nonionic form, and nonionized drug form is more ready to pass through the lipid cell membrane. Therefore, more nonionized drug at physiological pH will reach its target site more quickly and will have a faster onset of action. This explains why lignocaine has an enhanced onset of action than bupivacaine [69–72].

Local anesthetic is exposed to the more acidic axoplasmic side of the nerve when passed in the cell membrane. This leads to ionized form of the molecule which binds the sodium channel for blockade. Local anesthetics bind to serum α1-acid glycoproteins and other proteins. The duration of action for local anesthetics is based on high affinity against protein. As long as local anesthetic remains bound to nerve membranes, it provides an increased duration of action. Thus, the availability of free drug in the blood decreases which reduce the potential for toxicity.

Clinically, sodium bicarbonate is used to increase the pH of local anesthetic solutions. Nevertheless, it should be avoided from precipitation of anesthetic molecules over with alkalinization. Protein binding is also correlated to enhance continued blockade that alkaloid more firmly binds to the protein of the sodium channel. All local anesthetics are vasodilators (except cocaine) where the vasodilator activity leads to faster absorption and thus shorter duration of action. Addition of different alkaloids in local anesthetic solutions, i.e., mostly preferred epinephrine or norepinephrine and phenylephrine, produces vasoconstriction and provides less absorption through vascular beds. Thus, more local anesthetic will be available for neural blockade.

Local anesthetic sensitivity of nerve fibers differs to block generation and the conduction of nerve impulses. Small nerve fibers tend to be blocked more easily than larger fibers, and myelinated fibers are blocked sooner than non-myelinated fibers of the same diameter [22, 24, 40, 56, 57, 59, 73, 74].

6.2. Cocaine

Cocaine is a natural alkaloid produced from leaves of the coca that blocks nerve impulses and local vasoconstriction secondary to inhibition of local norepinephrine reuptake. However, it has a limited clinical use because of its toxicity and the potential for abuse. In modern medicine, cocaine is used primarily in topical anesthesia of the upper respiratory tract, where it's combined with vasoconstrictor and local anesthetic properties [56, 59, 75].

6.3. Lidocaine

Lidocaine was developed by Lofgren after procaine, the most widely used cocaine derivative with a low pKa (7.7). Lidocaine is a local anesthetic of the amide type as well as cardiac depressant (antiarrhythmic) which has an ingredient of lidocaine hydrochloride alkaloid ($C_{14}H_{22}N_2O{\cdot}HCl$). Lidocaine hydrochloride is a white powder freely soluble in water where injection of sterile, nonpyrogenic solution of it is indicated for production of local or regional anesthesia by infiltration techniques (percutaneous injection) and intravenous regional anesthesia by peripheral nerve block techniques (brachial plexus and intercostal) and by central neural techniques (lumbar and caudal epidural blocks). Further, it has been administered by continuous intra-articular infusion (to control postoperative pain). Lidocaine binds to neuronal membrane to stabilize and inhibit voltage-gated sodium channels for the initiation and conduction of impulses. Lidocaine has a very rapid onset of action that typically begins working within 4 min and an anesthetic half-life of about 1.6 h (30 min to 3 h). Therefore, lidocaine is suitable for infiltration, block, and surface anesthesia where its actions are more intense and its effects are more prolonged (when compared to procaine). Subdural and epidural anesthesias bupivacaine or prilocaine are preferred as longer-acting substances [37, 57, 58, 76].

The rapid metabolism by the liver enzymes leads to biotransformation of lidocaine in some pathways including oxidative N-dealkylation, ring hydroxylation, cleavage of the amide linkage, and conjugation. The pharmacologically active metabolites monoethylglycinexylidide and inactive glycinexylidide are produced as a result of N-dealkylation biotransformation that is a primary metabolism. Monoethylglycinexylidide has a longer half-life but also less potent than lidocaine. The rate of biotransformation of an amide local anesthetic significantly influences the anesthetic blood levels where increased blood levels can potentially increase toxicity. Approximately 90% of the dose of injected lidocaine undergoes biotransformation and is excreted in the form of various metabolites in the kidney, whereas less than 10% is excreted and unchanged.

The concentration of AAG and the concentration of free lidocaine affect the predictability of lidocaine toxicity where higher doses of lidocaine can be tolerated before encountering toxicity based on surgery type and certain disease states. For example, surgery, trauma, postoperative inflammation, cancer, and myocardial infarctions cause an increase in α1-acid glycoprotein and lidocaine binding, thus reducing the free lidocaine plasma concentration [77].

Through the lipid solubility, lidocaine penetrates in nerve membranes and binds to acute-phase protein of α1-acid glycoprotein (AAG) which is responsible for binding lidocaine and other basic drugs. Although lidocaine can bind to albumin (low affinity, high capacity), AAG (low capacity, high affinity) is the major drug-binding macromolecule even with lower plasma concentration. Albumin is considered as the principal binding protein for acidic compounds, and AAG is the principal binding protein for basic drugs [37, 57, 58, 76, 78, 79].

Cocaine is a natural alkaloid produced from leaves of the coca that blocks nerve impulses and local vasoconstriction secondary to inhibition of local norepinephrine reuptake. However, it has a limited clinical use because of its toxicity and the potential for abuse. In modern medicine, cocaine is used primarily in topical anesthesia of the upper respiratory tract, where it's combined with vasoconstrictor and local anesthetic properties [26].

6.4. Procaine

Procaine, the first synthetic derivative of cocaine, was created in 1904 with the trade name of Novocaine, from the Latin nov- (new) and -caine, a common ending for alkaloids used as anesthetics. Procaine, an amino ester, is a short-acting local anesthetic that is used as an injection for local infiltration and peripheral nerve block during surgery and other medical and dental procedures. Low potency, slow onset, and short duration of action are the characteristics of procaine. Although its clinical use is largely confined to infiltration anesthesia and peripheral nerve block as sterile solutions in concentrations of 1 and 2% for injection, this drug is no longer available in the United States.

Procaine has a pKa of 8.9 where commercially prepared solutions have a pH of 5.5 to 6.0. Due to rapid hydrolysis by plasma cholinesterase, it has very short plasma half-life that is thought to be approximately 20 seconds. It has a very short duration of action based upon extremely poor protein binding. While toxicity associated with the use of procaine is quite a little, the use of procaine produces metabolite para-aminobenzoic acid which is associated with an increased rate of allergy [57, 80].

6.5. Benzocaine

Benzocaine is an amino ester that is a derivative of procaine with a pKa of 8.5 and pH in preparation between 4.5 and 6.0. Benzocaine binds to sodium channels and reversibly depolarizes the neuronal membrane, consequently blocking the initiation and conduction of nerve impulses. It has a slow onset, short duration, and moderate toxicity. Benzocaine is considered an oral anesthetic because it numbs the mouth by dulling the nerve endings in painful areas [81].

6.6. Chloroprocaine

Chloroprocaine is a short-acting amino ester which is a chlorinated derivative of procaine. Because it is the most rapidly metabolized local anesthetic (by cholinesterase), it has an extremely short plasma half-life. Chloroprocaine has a pKa of 8.9, whereas prepared solutions have a pH of 2.5 to 4.0. Although it has a high pKa value, characterized by rapid onset of action, it also has the least toxicity in the central nerve system and/or cardiovascular system among current agents in use.

Chloroprocaine has clinical usage mostly in epidural anesthesia and also in peripheral blocks with short duration, whereas combined usage with other long-acting, slow-onset local anesthetics such as bupivacaine and tetracaine to achieve rapid onset with prolonged duration is common.

Owing to the short duration of spinal chloroprocaine, it is a strong alternative to lidocaine for surgical blocks and short or ultrashort surgical procedures for outpatient anesthesia. Its use in a dose ranging between 30 and 60 mg for procedures lasting >60 min has been suggested in the literature [82].

6.7. Tetracaine

Tetracaine is a long-acting amino ester with a pKa of 8.6, and commercially prepared solutions have a pH between 4.5 and 6.5. It has been used in spinal anesthesia if long duration of action is needed and also in various topical anesthetic preparations. When compared to the other commonly used ester local anesthetics, tetracaine is more slowly metabolized and considerably more toxic. It is significantly more potent and has a longer duration of action than procaine or chloroprocaine. Tetracaine is not recommended for peripheral nerve blocks owing to its slow onset and potential for systemic toxicity (may be greater if used with epinephrine). It has clinical use in spinal anesthesia with and without the use of epinephrine leading in a considerably reliable and long-onset spinal anesthetic [80].

6.8. Mepivacaine

Mepivacaine is an intermediate-duration amino amide local anesthetic with a pKa of 7.6. In terms of pharmacologic properties, mepivacaine is often compared to lidocaine where mepivacaine has similar onset of action with a slightly longer duration for infiltration anesthesia. When an intermediate-duration blockade is desired for peripheral nerve block techniques, 1.5% mepivacaine is an attractive local anesthetic, particularly in high-risk cardiac patients. Low toxicity, rapid onset, dense motor block, and excellent diffusion properties through tissue can be listed as characteristic of mepivacaine. Mepivacaine is clinically used as a local anesthetic for an epidural or spinal block and also used as an anesthetic for dental procedures. Some addictive people try to obtain this drug and inject it themselves because large doses of mepivacaine cause sedation, immobility, confusion, dissociation, and amnesia [83].

6.9. Bupivacaine

Bupivacaine is introduced in 1963 since then it has been one of the most commonly used long-acting local anesthetics in regional and infiltration anesthesia. The structural difference of bupivacaine from lidocaine is the amine-containing group that is a butylpiperidine. Bupivacaine is capable of producing prolonged anesthesia and analgesia, thus reducing the need for repeated administration and rarely required the addition of epinephrine. Epinephrine can be prolonged even further duration of blockade. Bupivacaine produces more sensory than motor block where anesthesia duration is from 4 to 16 h depending upon the site of injection and the concentration used. Bupivacaine stabilizes the neuronal membrane and is widely used both in neuraxial and peripheral nerve blockade. Bupivacaine hydrochloride has higher lipid solubility (logDpH 7.4 = 2.54) and a much decreased rate of hepatic degradation when compared to lidocaine. This characteristic makes bupivacaine more cardiotoxic than lidocaine. Because of its greater tendency to produce cardiotoxicity, large doses of bupivacaine should be avoided and is not recommended for intravenous regional analgesia. 0.5% bupivacaine is as effective as 2% lignocaine. Bupivacaine is widely used both in neuraxial and peripheral nerve blockade, infiltration anesthesia, spinal anesthesia, and epidural and caudal anesthesia. It is not recommended for intravenous regional analgesia because of cardiotoxicity which is much more than other local anesthetics. When bupivacaine is used, the smallest effective dose is aimed to administer [84–86].

6.10. Ropivacaine

The studies on reducing the cardiotoxicity of bupivacaine resulted to development of ropivacaine with similar physicochemical properties of onset, potency, and duration to those of bupivacaine. Ropivacaine is used in concentrations of 0.5% or higher and produces dense blockade with a slightly shorter duration than bupivacaine. The onset of blockade is almost as fast as 1.5% mepivacaine or 3% 2-chloroprocaine in concentrations of 0.75%. Having less cardiotoxicity and central nerve system toxicity, reduced motor block, and an absolute difference in potency are mentioned in the literature as the advantages of ropivacaine over bupivacaine. However, it can be suggested that there may be no more than slight differences in onset, but no difference between ropivacaine and bupivacaine in duration of block. For these reasons, both drugs have been used as an effective long-acting local anesthetic in peripheral nerve blockade [84, 87, 88].

6.11. Levobupivacaine

Despite the effectiveness of bupivacaine as anesthetic, there are safety concerns related to cardiovascular and/or central nerve system toxicity. As a result levobupivacaine is being associated with a lower risk of toxicity alternative to bupivacaine. Levobupivacaine contains a single enantiomer of bupivacaine hydrochloride (the S-(-)-enantiomer of (±)-bupivacaine) where dexbupivacaine (R-(+)-enantiomers) is the other enantiomer of bupivacaine. The route of administration and concentration effects the onset and duration of sensory and motor block where levobupivacaine, dexbupivacaine, or bupivacaine has similar potency as an anesthetic.

However, levobupivacaine is consistently less toxic than bupivacaine. The clinical use of levobupivacaine includes surgical anesthesia or pain management during labor, postoperative analgesia, lumbar epidural or intrathecal anesthesia, thoracic epidural anesthesia, peripheral nerve block, and infiltration anesthesia that have mostly investigated and compared with bupivacaine [84, 88].

6.12. Opiates

Opiate describes any of the narcotic opioid alkaloids found in opium plant (*Papaver somniferum*) that are morphine, codeine, thebaine, and papaverine or synthetic opioids that are derived from morphine and thebaine (oxycodone and hydrocodone). Opiates and its synthesized derivatives are prescribed in medications used for pain relief with even stronger analgesic properties than their predecessors do. Opium itself contains over 25 different alkaloids, whereas only morphine and codeine are used as opiate analgesics from the point of view of clinical significance. The majority of drugs appearing on a synthetic opioid list are derivatives of these medications. The list of opiate drugs includes natural opiate drugs from natural opium alkaloids and synthetic (synthetic and semisynthetic) opiate drugs, also called as opioids, made by chemical synthesis in the laboratory according to the chemical structures found in natural alkaloids.

Natural opiate drugs can be listed as morphine, codeine, thebaine, and oripavine. Synthesized versions of natural opiate drugs are Demerol, Fentanyl, Dilaudid, Norco, Lortab, Atarax, Methadone, and Buprenorphine. Semisynthetic medications are derived from the naturally occurring alkaloids where small concentrations of natural opium alkaloids go into the making of semisynthetic opiates, thus, a part of synthetic opiates. Semisynthetic opiate drugs include:

- Oxymorphone—contains the natural alkaloid, thebaine.
- Hydrocodone—contains the natural alkaloid, codeine.
- Oxycodone—contains the natural alkaloid, thebaine.
- Hydromorphone—contains the natural alkaloid, morphine.

The studies in the laboratory were focused on eliminating the risk for addiction of natural opioid alkaloids and make even more effective than opium itself for relatively safe alternatives for treatment. However, opiates in any form always carry a risk for addiction. Fully synthetic alkaloids are synthesized from other chemicals and molecules that do not come from alkaloids found in plants [61, 89].

Both of these natural and synthetic types of drugs bound to the active site on the receptors (called opioid receptors) from certain nerve cells in the brain, spinal cord, and gastrointestinal tract. Once the opioids injected into the epidural or subarachnoid space to manage acute or chronic pain, bound to the specific nerve cells in the brain, block a specific receptor site and sent inaccurate measures of the severity of the pain so that the person who has taken the drug will experiences less pain. Thus, opiate is used in combinations with a local anesthetic to both enhance the blockade and prolong analgesia in neuraxial blockade. Nevertheless, opioid receptors do not exist in peripheral nerve so opiates do not have an effective clinical role in peripheral nerve blockade. Taking drugs in these classes also affect how the brain feels pleasure, in other words addictive [21, 90–94].

Completely synthetic alkaloid molecules called opioids are originally synthesized by pharmaceutical companies because of their potency as an analgesic for the treatment of severe pain. As opposed to an opiate (natural opium alkaloid), they are very similar in structure to morphine, whereas the exact geometry of the molecule largely effects on determining the painkilling activity. For example, the drugs Levorphan and Dextrorphan are both mirror images of one another with quite similar structure to morphine. Levorphan, the left-handed molecule like naturally left-handed morphine, is several times more potent than morphine and is strongly addictive, while Dextrorphan, the right-handed molecule, has no analgesic ability and is also nonaddictive [21, 90–94].

6.12.1. Morphine

The main alkaloid morphine in opium poppy is a naturally occurring analgesic opioid and one of the most potent pain relievers. One of the most important uses of morphine in today's clinical practice is relief of pain caused by heart attack or myocardial infarction. Morphine helps to ease the pain particularly before, during, and after major surgeries and acts as an anesthetic without decreasing consciousness. Morphine is a weak base with the pKa of about 8.0 and has relatively low lipid solubility at physiological pH (**Figure 4**). Morphine exists mostly in ionized form that does not favor passage through the lipid membrane; thus, onset of action is relatively slow (15–30 min). The basic mechanism depends on the shape of the morphine molecule and its binding to the active site on the receptor protein. The ability of morphine to fit into the active site and block a specific receptor site on a nerve cell gives its analgesic properties. In this way the action of the pain receptor rules out and intercepts the pain signals reaching the brain. The specific opiate receptors are classified mainly as mu (μ), kappa (κ), and sigma (σ). In terms of analgesia effects, the mu (μ) is the most important receptor type including at least two subtypes, μ1 and μ2, in which μ1 probably mediates analgesia. Morphine produces relaxation and sedation by depressing the nervous system [91, 93, 95].

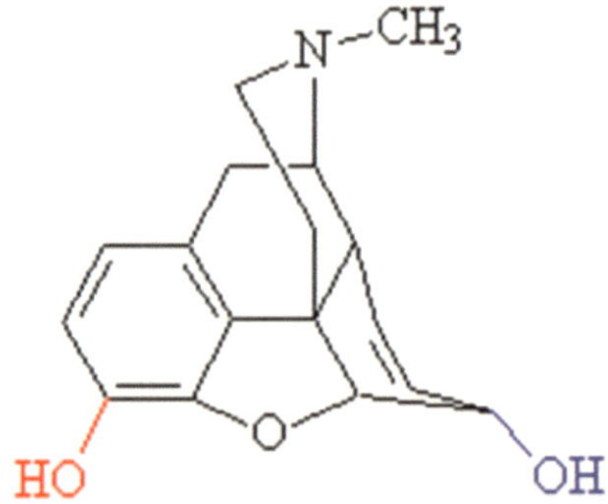

Figure 4. Morphine structure, replacing the —OH group shown in red with —OCH_3, produces codeine where heroin derives from replacing both the red and blue —OH groups with $OCOCH_3$.

http://dx.doi.org/10.5772/intechopen.69847

Hydromorphone is synthesized from morphine that is much stronger than morphine but has less side effects and a lower dependency rate. It is prescribed to treat severe pain or dry coughing (Dilaudid and Palladone).

Completely synthetic alkaloid molecules called opioids are originally synthesized by pharmaceutical companies because of their potency as an analgesic for the treatment of severe pain. As opposed to an opiate (natural opium alkaloid), they are very similar in structure to morphine, whereas the exact geometry of the molecule largely effects on determining the painkilling activity. For example, the drugs Levorphan and Dextrorphan are both mirror images of one another with quite similar structure to morphine. Levorphan, the left-handed molecule like naturally left-handed morphine, is several times more potent than morphine and is strongly addictive, while Dextrorphan, the right-handed molecule, has no analgesic ability and is also nonaddictive.

6.12.2. Codeine

Codeine is another powerful analgesic derived from morphine (although it is less potent than morphine) by replacing one of the —OH groups with a methoxy group (—OCH_3) and was first used as a therapeutic antitussive (see structure below in **Figure 5**). When codeine is metabolized, the —OCH_3 group converted back to —OH and regenerates morphine. Although morphine, heroine, and codeine are all derived from opium, codeine has a slight difference in chemical structure from morphine and heroine. It is often sold as a salt in the form of either codeine sulfate or codeine phosphate, while codeine hydrochloride is more common worldwide. In therapeutic doses, codeine is much less effective as an analgesic than morphine therefore often combined with other medications including Empirin with codeine, Fiorinal with codeine, and Tylenol with codeine. The prescription medications of codeine is available in pill form but also can be injected into the muscle. It is not given through IV as it can cause convulsions. Codeine is used for its analgesic, antitussive, antidiarrheal, antihypertensive, anxiolytic, antidepressant, sedative, and hypnotic properties [21, 96].

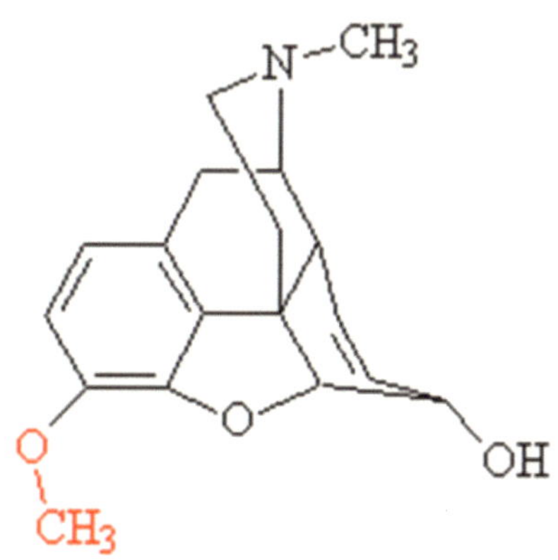

Figure 5. Replacing the —OH group of morphine shown in red with —OCH_3 produces codeine.

6.12.3. Heroin

Heroin was synthesized from morphine to be used as a nonaddictive cure; however, it was turned out that heroin is highly addictive and more potent than morphine. Heroin exhibits euphoric, anxiolytic, and analgesic central nervous system properties so classified as a Schedule I drug that possession or trading of it is illegal. Heroin (diacetylmorphine or diamorphine) is derived from the morphine alkaloid by acetylation of the two hydrogen-bonding hydroxyl groups (–OH) with acetyl chloride (–$OCOCH_3$) (**Figure 6**). Acetylation makes heroin more soluble in nonpolar solvents (i.e., oils and fats) and provides rapid pass through the blood-brain barrier which makes heroin much more potent than morphine because heroin is converted back to morphine in the brain. Patients with heroin addiction should seek advice from healthcare providers [91, 93, 95].

6.12.4. Thebaine

Thebaine is a natural alkaloid that is extracted from the opium poppy plant, due to its origin classified as opiates. Though thebaine is toxic and unsuitable for use as drug, pharmaceutical companies use thebaine alkaloids to create semisynthetic therapeutic drugs including Naloxone, Naltrexone, and Buprenorphine by breaking down it into smaller metabolites and also alter the chemical structure such as oxycodone. In spite thebaine binds to opioid receptors in the central nervous system, it stimulates the nervous system, while morphine and codeine produce relaxation and sedation by depressing the nervous system.

Thebaine is used to create semisynthetic pain relievers, such as oxymorphone and oxycodone. Oxycodone is synthesized from thebaine and used to treat moderate to severe pain for patients after surgery. Oxycodone is usually prescribed in tablet forms (Percocet, Percodan, or OxyContin) and widely combined with other drugs such as aspirin. Hydrocodone is synthesized from either codeine or thebaine and prescribed to treat moderate to severe pain and heavy coughs (Vicodin, Lorcet, Dolorex, Forten, and Anexsia).

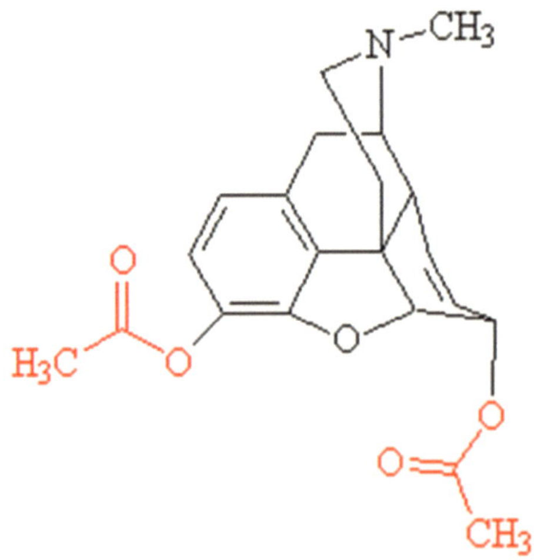

Figure 6. Chemical structure of heroin derived from morphine.

6.12.5. Oripavine

Oripavine is not produced by traditionally cultivated varieties of opium poppy where a chain of chemical processes are used to extract oripavine from the opium poppy plant. First of all opium is broken to major alkaloid metabolites, codeine, morphine, and thebaine. Oripavine is the major metabolite of thebaine that is produced by altering the chemical structure of thebaine for more effective or safer medications. Despite its development based on high therapeutic value and low-abuse potential compounds, it has severe toxicity and low therapeutic index. Even though oripavine is a strong pain reliever with comparable analgesic potency to morphine, it does not have clinical use. It is used as convenient source for the production of several synthetic opioid pharmaceuticals [97].

6.12.6. Fentanyl

Fentanyl is one of the most lipophilic opioids with rapid onset of action (10–15 min) and short duration (2–5 h) that found a wide usage epidural and intradural for postoperative pain. The clinical studies also focused on the beneficial effect of the combination of intrathecal fentanyl as lipophilic opioids with intrathecal local anesthetic in ambulatory surgery. Intrathecal bupivacaine is commonly preferred for local anesthesia in ambulatory surgery due to neurological toxicity of lidocaine. The fact remains that a high level of the sympathetic block can take place in high doses of intrathecal bupivacaine. The combination of intrathecal opioids added to low-dose local anesthetics provides a faster onset of blockade and better intraoperative and immediate postoperative analgesia that prevents increasing the degree of motor blockade or delaying discharge [98–100].

6.13. Other anesthetic alkaloids

Tropane alkaloids are synthesized from the amino acids ornithine, putrescine (decarboxylated ornithine), and proline or obtained by extraction from plants. The important drugs and alkaloids in this group are atropine, hyoscyamine, and scopolamine. These alkaloids have effects on the central nervous system, including nerve cells of the brain and spinal cord that may also affect the autonomic nervous system. The chemical structure of these includes a methylated nitrogen atom $N{-}CH_3$ at one end of the molecule which is also found in the neurotransmitter acetylcholine. Hence, they are capable of blocking or inhibiting nerve impulses between nerves in the brain and neuromuscular junctions. Atropine is an alkaloid which stimulates the central nervous system. Its sulfate salt is injected intramuscularly prior to induction of anesthesia. Reduction in pulse range and cessation of cardiac action are attributable to increased vigil activity administered intravenously during surgery. Scopolamine or hyoscine is an alkaloid that has a depressant activity on the central nervous system. Thus, scopolamine hydrobromide is used for preanesthetic sedation in conjunction with analgesics. Diterpenoid alkaloids (DAs) isolated from *Delphinium* and *Aconitum* plant species possess local anesthetic activity by suppressing sodium currents of excited membranes [22, 59, 101, 102].

6.14. Alkaloids for pain and pleasure

Despite alkaloids and other psychoactive chemicals came from nature act as defensive molecules for the organisms that made them, these types of molecules can be quite addictive drugs to humans. Many alkaloid substances like cocaine, caffeine, nicotine, opium (morphine, heroine), etc. act on the peripheral nervous system with the psychological side effects that produce a state of relaxed, mental excitement, and euphoria which are referred to addiction.

Medically useful alkaloids morphine and codeine are the pain relievers but also addictive drugs. Morphine is stronger than codeine, while codeine is often prescribed for moderate pain and also an effective cough suppressant. Heroin is another abused alkaloid drug derived from morphine by a simple chemical modification. However, heroin is metabolized to morphine once it enters the body; thus, these two alkaloids drugs are considered completely equivalent.

Cocaine is produced from the coca plant and classified as narcotic. The effects of cocaine can be sorted, excited, and elated, an impression of enhanced physical strength and mental ability feelings. There after elevated heart rate and blood pressure are accompanied to these feelings. However, cocaine produces a state of euphoric hyperarousal for a short duration; hence, the addicted person needs high dose of the drug that may lead to fibrillation and death. Cocaine usage over time can result in paranoid schizophrenia.

Nicotine is another alkaloid that is pleasurable and addictive mainly known for tobacco plants, whereas potatoes, eggplant, cauliflower, and tomato can be listed as nicotine sources. Nicotine alkaloids have primarily stimulation properties and produce either relaxation or arousal.

There are other alkaloids which are not addictive, but stimulants such as caffeine and the analog compounds theophylline and theobromine are. Caffeine occurs naturally in coffee, tea, cocoa and chocolate, cola drinks, and a variety of other plants. Tea contains small amounts of theophylline apart from caffeine, while theobromine is the major stimulant in cocoa. Cocoa and chocolate contain neuroactive alkaloids known as tetrahydro-beta-carbolines that possibly have influences on mood and behavior [23, 24, 103–108].

Author details

Havva Sayhan[1*], Serbülent Gökhan Beyaz[2] and Ahmet Çeliktaş[3]

*Address all correspondence to: hsayhan@gmail.com

1 Sakarya University Training and Research Hospital, Sakarya, Turkey

2 Sakarya University Training and Research Hospital, Korucuk Campus, Department of Anesthesiology and Pain Medicine, Sakarya, TURKEY

3 Tubitak Bilgem Yital, Kocaeli, TURKEY

References

[1] Konrath EL, Passos CS, Klein-Junior LC, Amélia TH. Alkaloids as a source of potential anticholinesterase inhibitors for the treatment of Alzheimer's disease. Journal of Pharmacy and Pharmacology. 2013;**65**(12):1701-1725. DOI: 10.1111/jphp.12090

[2] Liu J, Zhu Y, Yu Z, Zhou Y, Han Y, Wu F, Qi B. The structures of huperzine A and B, two new alkaloids exhibiting marked anticholinesterase activity. Canadian Journal of Chemistry. 2011;**64**(4):837-839. DOI: 10.1139/v86-137

[3] T.B. Ng, Liu F, Wang ZT. Antioxidative activity of natural products from plants. Life Sciences. 2000;**66**(8):709-723. DOI: 10.1016/S0024-3205(99)00642-6

[4] Liu M, Huang HH, Yang J, Su YP, Lin HW, Lin LQ, Liao WJ, Yu CX, The active alkaloids of *Gelsemium elegans* Benth. are potent anxiolytics. Psychopharmacology. 2013;**225**(4):839-851. DOI: 10.1007/s00213-012-2867-x

[5] Cowan MM. Plant products as antimicrobial agents. Clinical Microbiology Reviews. 1999;**12**(4):564-582

[6] Cos P, Maes L, Vlietinck A, Pieters L. Plant-derived leading compounds for chemotherapy of human immunodeficiency virus (HIV) infection – An update (1998–2007). Planta Medica. 2008;**74**(11):1323-1337. DOI: 10.1055/s-2008-1081314

[7] Wink M. Medicinal plants: A source of anti-parasitic secondary metabolites. Molecules. 2012;**17**(11):12771-12791. DOI: 10.3390/molecules171112771

[8] Kupeli E, Kosar M, Yesilada E, Husnu K, Baser KHC. A comparative study on the anti-inflammatory, antinociceptive and antipyretic effects of isoquinoline alkaloids from the roots of Turkish Berberis species. Life Sciences. 2002;**72**(6):645-657. DOI: 10.1016/S0024-3205(02)02200-2

[9] Kulkarni SK, Dhir A. Berberine: A plant alkaloid with therapeutic potential for central nervous system disorders. Phytotherapy Research. 2009;**24**(3):317-324. DOI: 10.1002/ptr.2968

[10] Imperatore C, Aiello A, D'Aniello F, Senese M, Menna M. Alkaloids from marine invertebrates as important leads for anticancer drugs discovery and development. Molecules. 2014;**19**:20391-20423. DOI: 10.3390/molecules191220391

[11] Aniszewski T. Alkaloids: Chemistry, Biology, Ecology, and Applications. 2nd ed. Elsevier Science; 2015. 496 p. ISBN: 978-0-444-59433-4

[12] Wink M. Allelochemical properties and the raison d'être of alkaloids. Alkaloids. 1993;**43**:1-118. Google Scholar

[13] Wink M, Latz-Brüning B, Schmeller T. Biochemical effects of allelopathic alkaloids. In: Inderjit KMMD, Chester LF, editors. Principles and Practices in Plant Ecology. Boca Raton, FL, USA: CRC Press; 1999. pp. 411-422. ISBN: 9780849321160

[14] Wink M. Interference of alkaloids with neuroreceptors and ion channels. Studies in Natural Products Chemistry. 2000;**21**:3-129. DOI: 10.1016/S1572-5995(00)80004-6

[15] Wink M. Molecular modes of action of cytotoxic alkaloids: From DNA intercalation, spindle poisoning, topoisomerase inhibition to apoptosis and multiple drug resistance. The Alkaloids: Chemistry and Biology. 2007;**64**:1-47. DOI: 10.1016/S1099-4831(07)64001-2

[16] Jung Jang E, Kil YS, Ryeon Park H, Oh S, Kyeong Kim H, Gyeong Jeong M, Kyoung Seo E, Sook Hwang E. Suppression of IL-2 production and proliferation of CD4(+) T cells by tuberostemonine. O. Chemistry & Biodiversity. 2014;**11**:1954-1962. DOI: 10.1002/cbdv.201400074

[17] Haznedaroglu MZ, Gokce G. Comparison of anti-acetylcholinesterase activity of bulb and leaf extracts of *Sternbergia candida* Mathew & T. Baytop. Acta Biologica Hungarica. 2014;**65**:396-404. DOI: 10.1556/ABiol.65.2014.4.4

[18] Eid SY, El-Readi MZ, Wink M. Digitonin synergistically enhances the cytotoxicity of plant secondary metabolites in cancer cells. Phytomedicine. 2014;**19**:1307-1314. DOI: 10.1016/j.phymed.2012.09.002

[19] Merschjohann K, Sporer F, Steverding D, Wink M, In vitro effect of alkaloids on bloodstream forms of *Trypanosoma brucei* and *T. congolense*. Planta Medica. 2001;**67**:623-627. DOI: 10.1055/s-2001-17351

[20] Phillipson JD. Phytochemistry and pharmacognosy. Phytochemistry. 2007;**68**(22-24):2960-2972. DOI: 10.1016/j.phytochem.2007.06.028

[21] Novak BH, Hudlicky T, Reed JW, Mulzer J, Trauner D. Morphine synthesis and biosynthesis—An update. Current Organic Chemistry. 2000;**4**(3):343-362. DOI: 10.2174/138527 2003376292

[22] Margaret FR, Michael W, editors. Alkaloids: Biochemistry, Ecology, and Medicinal Applications. 1st ed. New York: Plenum Press; 1998. 482 p. ISBN: 978-4757-2905-4

[23] Tadeusz A, editor. Alkaloids — Secrets of life. 1st ed. Amsterdam: Elsevier; 2007. 334 p. ISBN: 978-0444527363

[24] Hesse M, Alkaloids: Nature's curse or blessing?. 1st ed. Germany: Wiley-VCH; 2002. 413 p. ISBN: 3906390241, 9783906390246

[25] Dias DA, Urban S, Roessne U. A historical overview of natural products in drug discovery. Metabolites. 2012;**2**(2):303-336. DOI: 10.3390/metabo2020303

[26] Atanasov AG, Waltenberger B, Pferschy-Wenzig EM, Linder T, Wawrosch C, Uhrin P, Temml V, Wang L, Schwaiger S, Heiss EH, Rollinger JM, Schuster D, Breuss JM, Bochkov V, Mihovilovic MD, Kopp B, Bauer R, Dirsch VM, Stuppner H. Discovery and resupply of pharmacologically active plant-derived natural products: A review. Biotechnology Advances. 2015;**33**(8):1582-1614. DOI: 10.1016/j.biotechadv.2015.08.001

[27] Gurib-Fakim A. Medicinal plants: Traditions of yesterday and drugs of tomorrow. Molecular Aspects of Medicine. 2006;**27**(1):1-93. DOI: 10.1016/j.mam.2005.07.008

[28] Petrovska BB. Historical review of medicinal plants' usage. Pharmacognosy Reviews. 2012;**6**(11):1-5. DOI: 10.4103/0973-7847.95849

[29] East West School of Planetary Herbology. The Herbal Tradition [Internet]. 2017. Available from: http://www.planetherbs.com/history/the-herbal-tradition.html [Accessed: January 18, 2017]

[30] Amirkia V, Heinrich M. Alkaloids as drug leads – A predictive structural and biodiversity-based analysis. Phytochemistry Letters. 2014;**10**:48-53. DOI: 10.1016/j.phytol.2014.06.015

[31] Rates SMK. Plants as source of drugs. Toxicon. 2001;**39**(5):603-613. DOI: 10.1016/S0041-0101(00)00154-9

[32] Croteau R, Kutchan TM, Lewis NG. Natural products (secondary metabolites). In: Buchanan B, Gruissem W, Jones RL, editors. Biochemistry & Molecular Biology of Plants. 1st ed. American Society of Plant Physiologists, Chichester: Wiley; 2000. pp. 1250-1318. ISBN-13: 978-0943088396

[33] Cordell GA, Quinn-Beattie ML, Farnsworth NR. The potential of alkaloids in drug discovery. Phytotherapy Research. 2001;**15**(3):183-205. DOI: 10.1002/ptr.890

[34] Lesch JE. Conceptual change in an empirical science: The discovery of the first alkaloids. Historical Studies in the Physical Sciences. 1981;**11**(2):305-328. DOI: 10.2307/27757482

[35] Hagel JM, Facchini PJ, Benzylisoquinoline alkaloid metabolism: A century of discovery and a brave new world. Plant and Cell Physiology. 2013;**54**(5):647-672. DOI: 10.1093/pcp/pct020

[36] Hosztafi S, Fürst Z, Endogenous morphine. Pharmacological Research. 1995;**32**(1-2):15-20. DOI: 10.1016/S1043-6618(95)80003-4

[37] Wildsmith T. History and development of local anaesthesia. In: McLeod G, McCartney C, Wildsmith T, editors. Principles and Practice of Regional Anaesthesia. 4 ed. Oxford: Oxford University Press. 2012. DOI: 10.1093/med/9780199586691.003.0001

[38] Kennedy DO, Wightman EL. Herbal extracts and phytochemicals: Plant secondary metabolites and the enhancement of human brain function. Advances in Nutrition. 2011;**2**:32-50. DOI: 10.3945/an.110.000117

[39] O'Hagan D. Pyrrole, pyrrolidine, pyridine, piperidine and tropane alkaloids. Natural Products Reports. 2000;**17**:435-446. DOI: 10.1039/A707613D

[40] Fattorusso E, Taglialatela-Scafati O, editors. Modern Alkaloids: Structure, Isolation, Synthesis and Biology. 1st ed. Weinheim: Wiley; 2007. 690 p. ISBN: 978-3-527-31521-5

[41] Dewick PM, Medicinal Natural Products A Biosynthetic Approach. 2nd ed. Chichester, England: Wiley; 2001. 520 p. ISBN: 978-0471496403

[42] Halliwell B, Oxidative stress and cancer: Have we moved forward? Biochemical Journal. 2007;**401**(1):1-11. DOI: 10.1042/BJ20061131

[43] Chaves SK, Feitosa CM, Araújo LS. Alkaloids pharmacological activities - prospects for the development of phytopharmaceuticals for neurodegenerative diseases. Current Pharmaceutical Biotechnology. 2016;**17**(7):629-635. DOI: 10.2174/138920101707160503201541

[44] Pérez-Hernández J, Zaldívar-Machorro VJ, Villanueva-Porras D, Vega-Ávila E, Chavarría A. A potential alternative against neurodegenerative diseases: Phytodrugs. Oxidative Medicine and Cellular Longevity. 2016;**2016**:8378613. DOI: 10.1155/2016/8378613

[45] Zangara A. The psychopharmacology of huperzine A: An alkaloid with cognitive enhancing and neuroprotective properties of interest in the treatment of Alzheimer's disease. Pharmacology Biochemistry and Behavior. 2003;**75**(3):675-686. DOI: 10.1016/S0091-3057(03)00111-4

[46] Reisinea T, Bellb GI. Molecular biology of opioid receptors. Trends in Neurosciences. 1993;**16**(12):506-510. DOI: 10.1016/0166-2236(93)90194-Q

[47] Minami M, Satoh M. Molecular biology of the opioid receptors: Structures, functions and distributions. Neuroscience Research. 1995;**23**(2):121-145. DOI: 10.1016/0168-0102(95)00933-K

[48] Cushnie TPT, Cushnie B, Lamb AJ. Alkaloids: An overview of their antibacterial, antibiotic-enhancing and antivirulence activities. International Journal of Antimicrobial Agents. 2014;**44**(5):377-386. DOI: 10.1016/j.ijantimicag.2014.06.001

[49] Miliauskasa G, Venskutonisa PR, van Beek TA. Screening of radical scavenging activity of some medicinal and aromatic plant extracts. Food Chemistry. 2004;**85**(2): 231-237. DOI: 10.1016/j.foodchem.2003.05.007

[50] Bennett RN, Wallsgrove RM. Secondary metabolites in plant defence mechanisms. New Phytologist. 1994;**127**:617-633. DOI: 10.1111/j.1469-8137.1994.tb02968.x

[51] Glenn WS, Runguphan W, O'Connor SE. Recent progress in the metabolic engineering of alkaloids in plant systems. Current Opinion in Biotechnology. 2013;**24**(2):354-365. DOI: 10.1016/j.copbio.2012.08.003

[52] Bernhoft A, editor. Bioactive compounds in plants – Benefits and risks for man and animals. 1st ed. Oslo, Norway: AIT Otta AS; 2010. 253 p. ISBN: 978-82-7099-583-7

[53] Azwanida NN. A review on the extraction methods use in medicinal plants, principle, strength and limitation. Medicinal & Aromatic Plants. 2015;**4**:196. DOI: 10.4172/2167-0412.1000196

[54] Shinji F, Geoffrey AC. Alkaloids: A Treasury of Poisons and Medicines. 1st ed. London: Elsevier; 2014. 294 p. ISBN: 978-0124173026

[55] Lu JJ, Bao JL, Chen XP, Huang M, Wang YT. Alkaloids isolated from natural herbs as the anticancer agents. Evidence-Based Complementary and Alternative Medicine. 2012;**2012**:485042. DOI: 10.1155/2012/485042

[56] Becker DE, Reed KL. Essentials of local anesthetic pharmacology. Anesthesia Progress. 2006;**53**(3):98-109. DOI: 10.2344/0003-3006(2006)53[98:EOLAP]2.0.CO;2

[57] Skinner D, Swain A, Peyton R, Robertson C, editors. Analgesia and anaesthesia. Cambridge Textbook of Accident and Emergency Medicine. 1st ed. Cambridge, UK: Cambridge University Press; 1997. 1051 p. ISBN: 9780521433792

[58] Landau R, Basil Achilladelis B, Scriabine A, editors. Discovery and Development of Major Drugs Currently in Use in Pharmaceutical Innovation: Revolutionizing Human Health. 1st ed. Philadelphia: Chemical Heritage Press; 1999. 408 p. ISBN: 9780941901215

[59] Bruce NC. Alkaloids. In: Rehm HJ, Reed G, editors. Biotechnology Set. 2nd ed. Weinheim, Germany: Wiley-VCH Verlag GmbH; 2001. ISBN: 9780941901215

[60] Qiu S, Zhang A, Yan G, Wang X, Natural alkaloids: Basic aspects, biological roles, and future perspectives. Chinese Journal of Natural Medicines. 2014;**12**(6):401-406. DOI: 10.1016/S1875-5364(14)60063-7

[61] Katzung BG, Masters SB, Trevor AJ, editors. Basic and Clinical Pharmacology, Opioid Analgesics & Antagonists (Chapter 31). 12th ed. Lange: McGraw-Hill; 2012. 1248 p. ISBN: 978-0071764018

[62] Covino BG, Pharmacology of local-anesthetic agents. British Journal of Anaesthesia. 1986;**58**(7):701-716. DOI: 10.1093/bja/58.7.701

[63] Ruetsch YA, Boni T, Borgeat A. From cocaine to ropivacaine: The history of local anesthetic drugs. Current Topics in Medicinal Chemistry. 2001;**1**(3):175-182. DOI: 10.2174/1568026013395335

[64] Griffin WJ, Lin GD. Chemotaxonomy and geographical distribution of tropane alkaloids. Phytochemistry. 2000;**53**(6):623-637. DOI: 10.1016/S0031-9422(99)00475-6

[65] Snyder SH, Pasternak GW. Historical review: Opioid receptors. Trends in Pharmacological Sciences. 2003;**24**(4):198-205. DOI: 10.1016/S0165-6147(03)00066-X

[66] Scholz A. Mechanisms of (local) anaesthetics on voltage-gated sodium and other ion channels. British Journal of Anaesthesia. 2002;**89**(1):52-61. DOI: 10.1093/bja/aef163

[67] Fozzard HA, Lee PJ, Lipkind GM. Mechanism of local anesthetic drug action on voltage-gated sodium channels. Current Pharmaceutical Design. 2005;**11**(21):2671-2686. DOI: 10.2174/1381612054546833

[68] Narahashi T. Neuroreceptors and ion channels as the basis for drug action: Past, present, and future. Journal of Pharmacology and Experimental Therapeutics. 2000;**294**(1):1-26

[69] Tucker GT. Pharmacology of local anaesthetics. British Journal of Anaesthesia. 1986;**58**:701-716

[70] Mazoit JX, Dalens BJ. Pharmacokinetics of local anaesthetics in infants and children. Clinical Pharmacokinetics. 2004;**43**(1):17-32. DOI: 10.2165/00003088-200443010-00002

[71] Heavner JE. Local anesthetics. Current Opinions Anaesthesiology. 2007;**20**:336-342

[72] Nysora, Gadsden J. Local Anesthetics: Clinical Pharmacology and Rational Selection [Internet]. Available from: http://www.nysora.com/regional-anesthesia/foundations-of-ra/3492-local-anesthetics-clinical-pharmacology-and-rational-selection.html [Accessed: December 23, 2016]

[73] Strichartz GR, Ritchie JM. The action of local anesthetics on ion channels of excitable tissues. In: Strichartz GR, editor. Local Anesthetics. 1st ed. Berlin: Springer Berlin Heidelberg; 1987. pp. 21-52. ISBN: 978-3-642-71112-1. DOI: 10/1007/978-642-71110-7

[74] Dickenson AH. Where and how do opioids act? Proceedings of the 7th World Congress on Pain. In: Gebhart GF, Hammond DL, Jensen TS, editors. Progress in Pain Research and Management, Vol. 2. Seattle: IASP Press; 1994. pp. 525-552

[75] Calatayud J, González A. History of the development and evolution of local anesthesia since the coca leaf. Anesthesiology. 2003;**98**:1503-1508. DOI: 10.1097/00000542-200306000-00031

[76] Israili ZH, Dayton PG. Human alpha-1-glycoprotein and its interactions with drugs. Drug Metabolism Reviews. 2001;**33**(2):161-235. DOI: 10.1081/DMR-100104402

[77] Pieper JA, Slaughter RL, Anderson GD, Wyman MG, Lalka D. Lidocaine clinical pharmacokinetics. Annals of Pharmacotherapy. 2016;**16**(4):291-294. DOI: 10.1177/106002808 201600403

[78] Gordh T, Gordh TE, Lindqvist K. Lidocaine: The origin of a modern local anesthetic. Anesthesiology. 2010;**113**:1433-1437. DOI: 10.1097/ALN.0b013e3181fcef48

[79] Haas DA, An update on local anesthetics in dentistry. Journal of the Canadian Dental Association. 2002;**68**(9):546-551

[80] Langerman L, Grant GJ, Zakowski M, Golomb E, Ramanathan S, Turndorf H. Prolongation of epidural anesthesia using a lipid drug carrier with procaine, lidocaine, and tetracaine. Anesthesia & Analgesia. 1992;**75**(6):900-905

[81] Tetzlaff JE. Amino ester local anesthetics. In: Medicinal Chemistry, Edward B. Walker's Chemistry [Internet]. Weber State University. Available at: http://faculty.weber.edu/ewalker/Medicinal_Chemistry/topics/Psycho/local_a_ester.htm [Accessed: February 20, 2017]

[82] Ghisi D, Bonarelli S. Ambulatory surgery with chloroprocaine spinal anesthesia: A review. Dove Medical Press. 2015;**2015**(2):111-120. DOI: 10.2147/AA.S64884

[83] Brockmann WG. Mepivacaine: A closer look at its properties and current utility. General Dentistry. 2014;**62**(6):70-75

[84] Foster RH, Markham A. Levobupivacaine a review of its pharmacology and use as a local anaesthetic. Drugs. 2000;**59**(3):551-579. DOI: 10.2165/00003495-200059030-00013

[85] Knudsen K, Suurküla BM, Blomberg S, Sjövall J, Edvardsson SN. Central nervous and cardiovascular effects of i.v. infusions of ropivacaine, bupivacaine and placebo in volunteers. British Journal of Anaesthesia. 1997;**78**(5):507-514. DOI: 10.1093/bja/78.5.507

[86] Clarkson CW, Hondeghem LM. Mechanism for bupivacaine depression of cardiac conduction: Fast block of sodium channels during the action potential with

slow recovery from block during diastole. Anesthesiology. 1985;**62**(4):396-405. DOI: 10.1097/00000542-198504000-00006

[87] Whiteside JB, Wildsmith JAW. Developments in local anaesthetic drugs. British Journal of Anaesthesia. 2001;**87**(1):27-35. DOI: 10.1093/bja/87.1.27

[88] Shigeo O, Masayuki K, Toshikazu O, Ken Y, Tsutomu K. Systemic toxicity and resuscitation in bupivacaine-, levobupivacaine-, or ropivacaine-infused rats. Anesthesia & Analgesia. 2001;**93**(3):743-748. DOI: 10.1097/00000539-200109000-00039

[89] Holden JE, Jeong Y, Forrest JM. The endogenous opioid system and clinical pain management. AACN Clinical Issues. 2005;**16**(3):291-301. DOI: 10.1097/00044067-200507000-00003

[90] Chahl LA. Opioids—Mechanisms of action. Experimental and Clinical Pharmacology. 1996;**19**:63-51. DOI: 10.18773/austprescr.1996.063

[91] Taber DF, Neubert TD, Schlecht MF. The Enantioselective synthesis of morphine. Strategies and Tactics in Organic Synthesis. 2004;**5**:353-389. DOI: 10.1016/S1874-6004(04)80034-0

[92] Rinner U, Hudlicky T. Synthesis of morphine alkaloids and derivatives. Topics in Current Chemistry. 2012;**309**:33-66. DOI: 10.1007/128_2011_133

[93] Morimoto S, Suemori K, Moriwaki J, Taura F, Tanaka H, Aso M, Tanaka M, Suemune H, Shimohigashi Y, Shoyama Y. Morphine metabolism in the opium poppy and its possible physiological function, biochemical characterization of the morphine metabolite, bismorphine. Journal of Biological Chemistry. 2001;**276**(41):38179-38184. DOI: 10.1074/jbc.M107105200

[94] Al-Hasani R, Bruchas MR. Molecular mechanisms of opioid receptor-dependent signaling and behavior. Anesthesiology. 2011;**115**(6):1363-1381. DOI: 10.1097/ALN.0b013e318238bba6

[95] Filliol D, Ghozland S, Chluba J, Martin M, Matthes HW, Simonin F, Befort K, Gaveriaux-Ruff C, Dierich A, LeMeur M, Valverde O, Maldonado R, Kiefer BL. Mice deficient for delta- and mu-opioid receptors exhibit opposing alterations of emotional responses. Nature Genetics. 2000;**25**(2):195-200. DOI: 10.1038/76061

[96] Eddy NB, Hans F, Klaus-Jorgen H, Hans H. Codeine and its alternates for pain and cough relief. 3. The antitussive action of codeine- mechanism, methodology and evaluation. Bulletin of the World Health Organization. 1969;**40**:425-454. PMCID: PMC2554640

[97] Murphy B, Snajdr I, Machara A, Endoma-Arias MAA., Stamatatos TC, Cox DP, Hudlick T. Conversion of thebaine to oripavine and other useful intermediates for the semisynthesis of opiate-derived agents: Synthesis of hydromorphone. Advanced Synthesis & Catalysis. 2014;**356**:2679-2687. DOI: 10.1002/adsc.201400445

[98] Hamber EA, Viscomi CM. Intrathecal lipophilic opioids as adjuncts to surgical spinal anesthesia. Regional Anesthesia and Pain Medicine. 1999;**24**:255-263. PubMed ID: 10338179

[99] Rathmell JP, Lair TR, Nauman B. The role of intrathecal drugs in the treatment of acute pain. Anestheasia and Analgesia. 2005;**101**:S30-S43. DOI: 10.1213/01.ANE.0000177101.99398.22

[100] Bujedo BM. Recommendations for spinal opioids clinical practice in the management of postoperative pain. Journal of Anesthesiology and Clinical Science. 2013;**2049**(9752):2-28. DOI: 10.7243/2049-9752-2-28

[101] Grynkiewicz G, Gadzikowska M. Tropane alkaloids as medicinally useful natural products and their synthetic derivatives as new drugs. Pharmacological Reports. 2008;**60**(4):439-463. PubMed ID: 18799813

[102] Turabekova MA, Rasulev BF, Dzhakhangirov FN, Toropov AA, Leszczynska D, Leszczynski J. Aconitum and delphinium diterpenoid alkaloids of local anesthetic activity: Comparative QSAR analysis based on GA-MLRA/PLS and optimal descriptors approach. Journal of Environmental Science and Health, Part C. 2014;**32**:213-238. DOI: 10.1080/10590501.2014.938886

[103] Rosenblum A, Marsch LA, Joseph H, Portenoy RK. Opioids and the treatment of chronic pain: Controversies, current status, and future directions. Experimental and Clinical Psychopharmacology. 2008;**16**(5):405-416. DOI: 10.1037/a0013628

[104] Drugs for Pleasure, Drugs for Pain? Developing Treatments with Controlled Drugs Part Two: Opium, Morphine, & Heroin. [Internet]. Museum of the Royal Pharmaceutical Society; 2011. Available from: http://studylib.net/doc/8111046/drugs-for-pleasure--drugs-for-pain%3F. [Accessed: January 12, 2017]

[105] Science.jrank.org. Alkaloid—Alkaloids for Pain and Pleasure [Internet]. 2017. Available from: http://science.jrank.org/pages/229/Alkaloid-Alkaloids-pain-pleasure.html#ixzz4TlfGGpLw [Accessed: January 12, 2017]

[106] Bolton S, Null G. Caffeine: Psychological effects, use and abuse. Orthomolecular Psychiatry. 1981;**10**(3):202-211

[107] Nehlig A, Daval JL, Debry G, Caffeine and the central nervous system: Mechanisms of action, biochemical, metabolic, and psychostimulant effects. Brain Research. Brain Research Reviews. 1992;**17**(2):139-170

[108] Hukkanen J, Jacob P, Benowitz NL, Metabolism and disposition kinetics of nicotine. Pharmacological Reviews. 2005;**57**(1):79-115. DOI: 10.1124/pr.57.1.3

Chapter 4

Sceletium Plant Species: Alkaloidal Components, Chemistry and Ethnopharmacology

Srinivas Patnala and Isadore Kanfer

Additional information is available at the end of the chapter

http://dx.doi.org/10.5772/66482

Abstract

The genus *Sceletium,* classified under the Aizoaceae family, is indigenous to the Western, Eastern and Northern Cape province of South Africa. There are currently eight reported species divided into two main "types" with five species in the *tortuosum* and three in the *emarcidum* type. It has been observed that, in general, mesembrine-type alkaloids such as mesembrenol, Δ^7mesembrenone, mesembranol, mesembrenone, mesembrine and epimesembranol as well as some non-mesembrine type such as Sceletium A4, tortuosamine and joubertiamine occur in the *tortuosum* type; the *emarcidum* type is devoid of alkaloids. Morphological identification of species type presents a formidable challenge, where subtle differences are found in the secondary veins that branch off from the middle vein toward the leaf margin. In view of the fact that the plant contains a complex mixture of closely related compounds, in particular alkaloidal components, separation techniques and their application to evaluate specific chemical components are an important aspect which permits accurate characterization and quantification. In addition, the development of appropriate analytical methods for chemotaxonomic studies has provided valuable information to confirm specific plant identity. Importantly, these methods are also required for the quality control of plant material used to manufacture complementary and traditional medicines containing *Sceletium*.

Keywords: *Sceletium tortuosum, Sceletium emarcidum,* alkaloids, mesembrine, chemotaxonomy, HPLC-MS

1. Introduction and background

Sceletium alkaloids have been studied over a century when their presence was first reported in1896 and later by Zwicky in 1914. In a detailed study by Zwicky on about 40 species of the genus *Mesembryanthemum,* more than 50% of the plants tested positive for alkaloids. Due to this large number of species, the genus *Mesembryanthemum* was abandoned and some of the species were reassigned to genus *Sceletium*, family Aizoaceae [1].

These alkaloids, originating from *Sceletium* plants species, were widely found in the Western and Karoo regions of South Africa. The name *Sceletium* is derived from the Latin word *Sceletus* meaning skeleton. The derivation of the name is due to the prominent lignified leaf vein structure that is observed in dried leaves of this genus which give a skeletal appearance. Anecdotal evidence suggests that this plant is highly revered and held in great esteem by the tribes who collected and bartered it frequently in exchange for cattle and other commodities. Subsequently, the early Dutch colonists further showed commercial interest in this plant, and many plants of this family were also introduced to European cultivation [2].

The *Sceletium* plants can readily be identified by its persistent dry "skeletonized" leaves which enclose the young leaves during the dry season (**Figure 1a**), to protect them from adverse environmental conditions [3]. The specimens of two main types of Sceletium plants: *Sceletium tortuosum* and *Sceletium emarcidum* are depicted in **Figure 1b** and **c**, respectively.

Figure 1. (a) Skeletonized leaves of *S. tortuosum*. (b) *Sceletium tortuosum*. (c) *Sceletium emarcidum* (with skeletonized leaves).

2. *Sceletium* species

Sceletium species occurs in the Eastern, Northern, Western Cape provinces of South Africa and the genus *Sceletium*, belongs to the family, Aizoaceae [1].

2.1. Identification of *Sceletium* plant species

The specimens were studied and identified using the identification key of Gerbaulet [1]. Based on the identification key, the venation pattern which differs between species is one of the important taxonomic identification features.

There are currently eight reported species [3] of this genus, divided into two "types" with five species in the *tortuosum* type and three in *emarcidum* type as follows:

Tortuosum type: *Sceletium tortuosum; Sceletium crassicaule; Sceletium strictum; Sceletium expansum and Sceletium varians.*

Emarcidum type: *Sceletium emarcidum; Sceletium exalatum and Sceletium rigidum.*

The main differences are found in the secondary veins that branch off from the middle vein toward the leaf margin. Based on the venation type, the species is mainly classified as either *emarcidum* or *tortuosum* types (**Figure 2**). In the *emarcidum* type, the leaf is more flat and the dried leaf venation pattern shows a central main vein with the curved secondary vein which branches off the main vein, reaching the leaf margins.

In plants of the *tortuosum* type (**Figure 2**), the dry leaves are more concave and usually show about three to five or sometimes up to seven major parallel veins. The secondary veins run straight up to the apex on both sides of the middle vein.

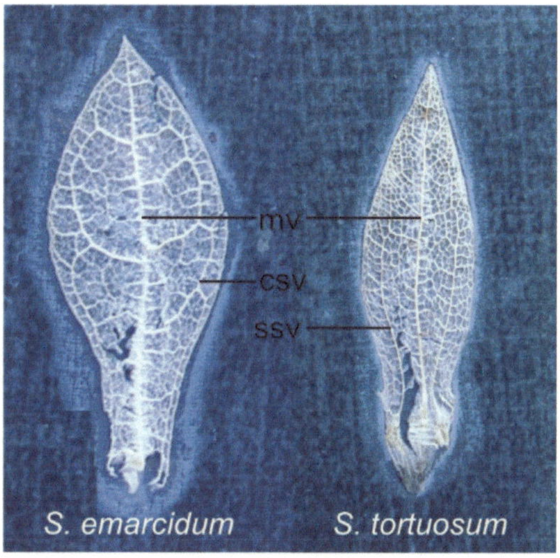

Figure 2. Venation pattern of skeletonized leaves in *Sceletium* species. mv = middle vein, csv = curved secondary vein, ssv = straight secondary vein.

3. Chemistry of *Sceletium* alkaloids

Preliminary studies on *Sceletium* were done by Meiring in 1896, suggesting that the presence of alkaloids and this was confirmed by Zwicky in 1914. Further studies on *S. expansum* and *S. tortuosum* reported by Zwicky in 1914, yielded a noncrystalline alkaloid which was named "mesembrin" with the reported molecular formula, $C_{16}H_{19}NO_4$ [4]. Rimington and Roets [5] reinvestigated this plant in 1937, and attempts to crystallize the alkaloid as a free base or hydrochloride salt were unsuccessful. In their experiments, they managed to obtain a crystalline picrate and platinichloride from the methylated free base and the molecular formula was deduced based on combustion analysis. The molecular formula for "mesembrin" was reassigned as $C_{17}H_{23}NO_3$ and is presently known as mesembrine, suggesting that the molecule belonged to the tropane ester alkaloid group.

Bodendorf and Krieger [6], in their work in 1957, revisited the molecule and successfully crystallized the mesembrine base to its hydrochloride salt, along with isolation of two more bases, namely "mesembrinine," presently known as mesembrenone, which has two hydrogen atoms less, and the structure is closely related to mesembrine. The other base was called "channaine," which was described as a phenolic base, and it was also reported that all these three compounds were purported to be optically inactive.

Popelak and Lettenbauer [7] in 1967 reported the incidence of *Sceletium* alkaloids in the plants they studied as 1 to 1.5%, which consisted of approximately 0.7% mesembrine and 0.2% "mesembrinine." The structure of mesembrine, deduced from their study, was reported as *N*-methyl-3a-(3′,4′-dimethoxyphenyl)-6-oxo-*cis*octahydroindole, which provided the foundation for continued studies on this group of alkaloids [4].

Jeffs *et al* in 1974 [8] worked further on *S. namaquense* and *S. strictum* and reported five new alkaloids, namely *Sceletium* alkaloid A4, N-formyltortuosamine, 4′-O-demethylmesembrenone, Δ^7mesembrenone and sceletenone. It was also reported that in a concurrent study by Wiechers *et al* on *S. tortuosum*, another base, tortuosamine, was isolated and had a close structural relation to *Sceletium* alkaloid A4.

Arndt and Kruger in 1970 [9] reported three new alkaloids, joubertiamine, dihydrojoubertiamine and dehydrojoubertiamine from *S. joubertii*, where their basic skeletons were biogenetically closely related to mesembrane (**Figure 3**) and not related to the mesembrine—like of alkaloids. The above alkaloids were also isolated and reported in another *Sceletium* species, *S. subvelutinum*, by Herbert and Kattah 1990 [10].

Whereas the phytochemical content of *Sceletium* species has been studied since 1896 [4], the reported alkaloidal content has been constrained to tortuosum-type species only, and related information on the emarcidum species has been conspicuously absent from the literature. However in 2013, Patnala and Kanfer reported the complete absence of mesembrine as well as other alkaloids usually found in the tortuosum type in their investigations involving three emarcidum species: *S. emarcidum, S. exalatum* and *S. rigidum* [11].

The alkaloids which have been isolated from *Sceletium* species are broadly classified into four structural classes. The major subgroup being the 3a-aryl-cis-octahydroindole skeleton which is referred to as the mesembrine group (**Table 1**) which includes Δ^4 series and Δ^7 series based on the double bond at position 4–5 (**Table 2**) and 7–7a (**Table 3**), respectively. *Sceletium* alkaloid A4 (**Table 4**) constitutes the lone member of the second subgroup. The third subgroup is closely related to the second, which is the alkaloid, tortuosamine type (**Table 5**), and the fourth group is the joubertiamine type (**Table 6**), which is closely related to the mesembrine series [10].

Of the above subgroups, the mesembrine type is the largest, consisting of about 15 alkaloids. The class derives its name from mesembrine, which was the first structurally characterized alkaloid molecule [4].

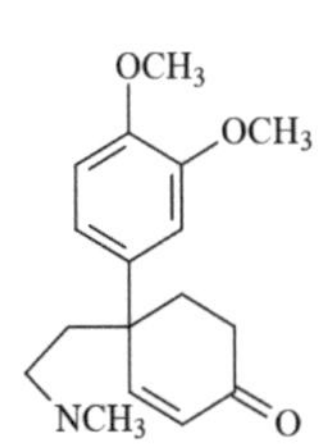

Figure 3. Mesembrane.

The major alkaloid in mesembrine type is (−)-mesembrine, reported to be present in up to 1% in *S. namaquence* and occurs as a partial racemate in *S. strictum* and *S. tortuosum* in smaller amounts [8]. The reported alkaloids in this subgroup are listed in **Tables 1–3** [12].

3.1. Mesembrine-type (I)

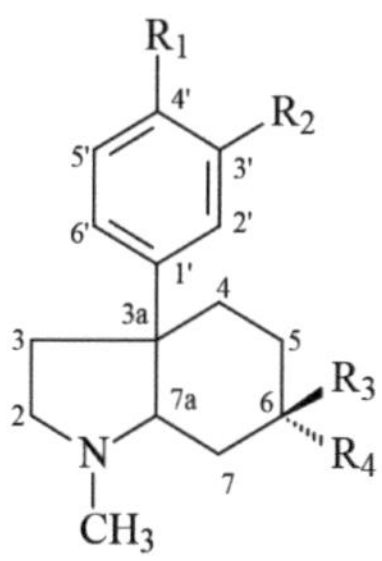

No.	R1	R2	R3, R4	R3	R4	Compound
1	OMe	OMe	O	–	–	Mesembrine
2	OMe	OMe	–	OH	H	Mesembranol
3	OMe	OMe	–	H	OH	Epimesembranol
4	OMe	OMe	–	OAc	H	Mesembranol acetate
5	OH	OMe	–	OH	H	4'Demethyl mesembranol
6	OMe	OMe	–	OMe	H	Mesembranol methyl ether
7	OMe	OMe	H	–	–	Mesembrane

Table 1. Mesembrine-type (I) *Sceletium* alkaloids.

3.2. Δ⁴ Mesembrine-type (II)

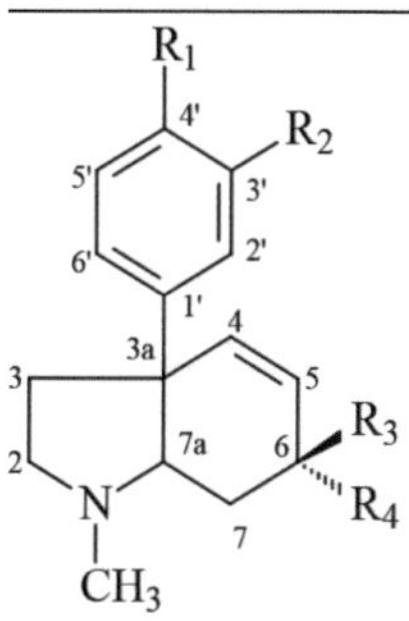

No.	R1	R2	R3,R4	R3	R4	Compound
8	OMe	OMe	O	–	–	Mesembrenone
9	OMe	OMe	–	OH	H	Mesembrenol
10	OMe	OMe	–	H	OH	6-Epimesembrenol
11	OMe	OMe	–	OAc	H	Mesembrenol acetate
12	OH	OMe	–	OMe	H	–
13	OMe	H	O	–	–	4'-O-methyl sceletenone
14	OH	H	O	–	–	Sceletenone

Table 2. Δ⁴Mesembrine-type (II) *Sceletium* alkaloids.

3.3. Δ⁷ Mesembrine-type (III)

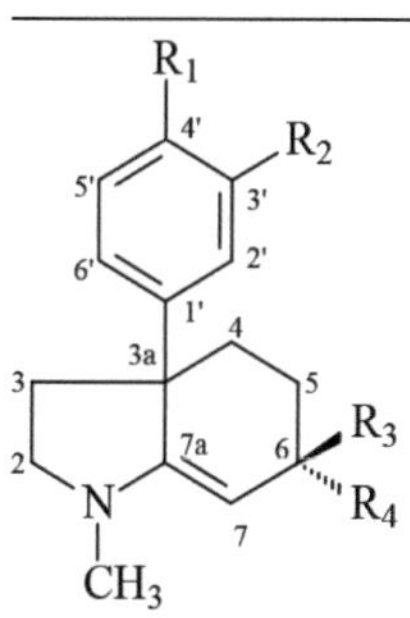

No.	R1	R2	R3,R4	R3	R4	Compound
15	OMe	OMe	O	–	–	Δ⁷ Mesembrenone

Table 3. Δ⁷Mesembrine-type (III) *Sceletium* alkaloid.

3.4. *Sceletium* A4 types (IV)

Table 4 depicts *Sceletium* A4 alkaloid (16) and is reported to occur in *S. namaquense* as an optically active crystalline base. The other reported alkaloid [8] which is closely related to this structure is a noncrystalline optically active compound mentioned as dihydropyridone base (17).

	No.	R1	R2	Compound
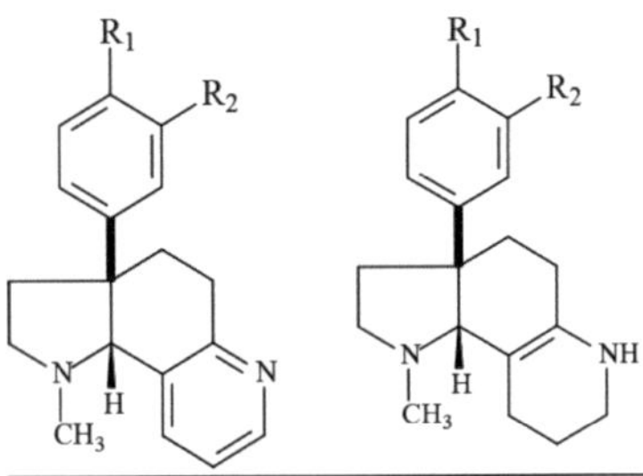	16	OMe	OMe	*Sceletium* A4
	17	OMe	OMe	Dihydropyridone base

Table 4. *Sceletium* A4 type (IV) alkaloids.

3.5. Tortuosamine type (V)

The reported alkaloids (**Table 5**) in this subclass are tortuosamine (18), N-formyltortuosamine (19) and N-acetyltortuosamine (20). Tortuosamine, a noncrystalline optically active base, was isolated from *S. tortuosum* [8].

	No.	R1	R2	R3	Compound
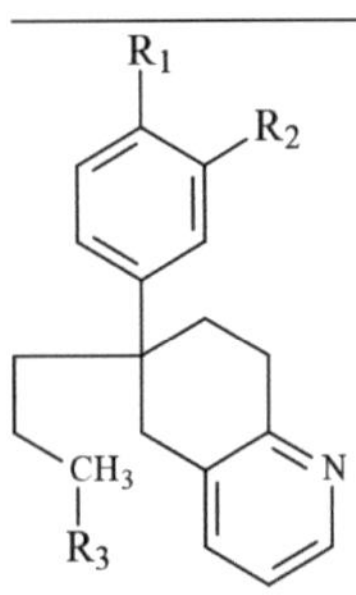	18	OMe	OMe	H	Tortuosamine
	19	OMe	OMe	CHO	N-formyltortuosamine
	20	OMe	OMe	COMe	N-acetyltortuosamine

Table 5. Tortuosamine-type (V) *Sceletium* alkaloids.

3.6. Joubertiamine types

These alkaloids are reported to occur principally in *S. joubertii* and have also been reported to occur in *S. subvelutinum*. These alkaloids are further classified as depicted in **Tables 6–8** [8].

http://dx.doi.org/10.5772/66482

3.6.1. Dihydrojoubertiamine (VI)

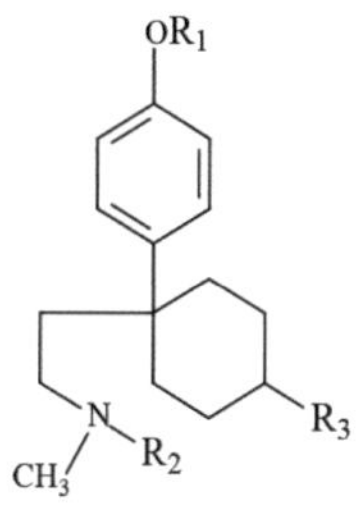

No.	R1	R2	R3	Compound
21	H	Me	O	dihydrojoubertiamine
22	H	Me	Me	O-methyldihydrojoubertiamine

Table 6. Dihydrojoubertiamine-type (VI) *Sceletium* alkaloids.

3.6.2. Dehydrojoubertiamine (VII)

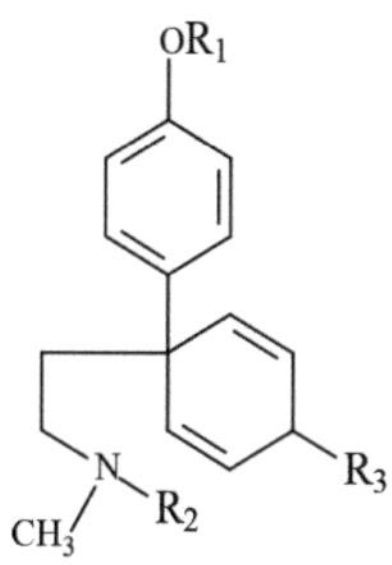

No.	R1	R2	R3	Compound
23	H	Me	O	dehydrojoubertriamine

Table 7. Dehydrojoubertiamine-type (VII) *Sceletium* alkaloid.

3.6.3. Joubertiamine (VIII)

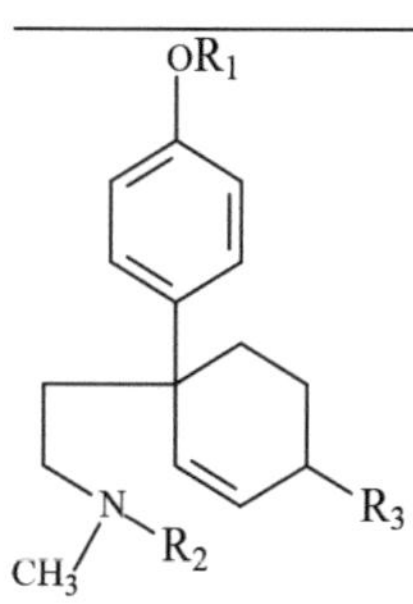

No.	R1	R2	R3	Compound
24	H	Me	O	Joubertiamine
25	Me	Me	O	O-methyljoubertiamine

Table 8. Joubertiamine-type (VIII) *Sceletium* alkaloids.

Structure[21]	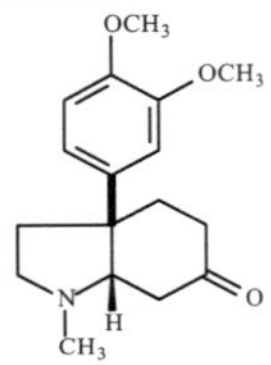	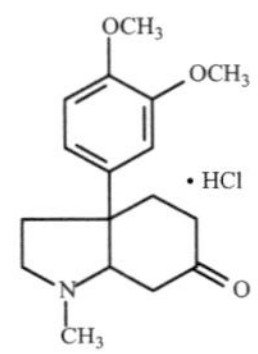	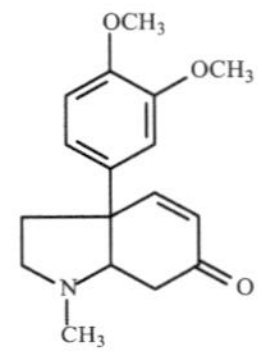	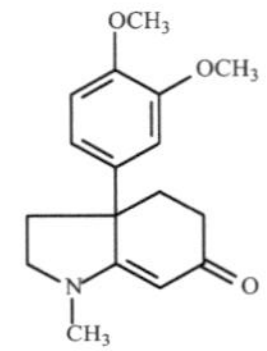	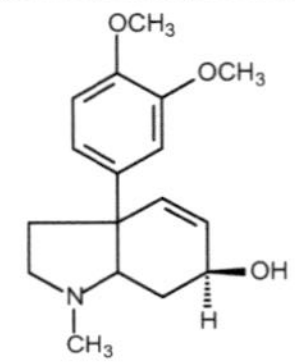	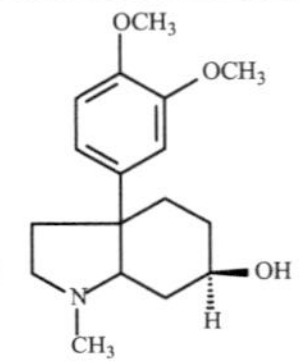	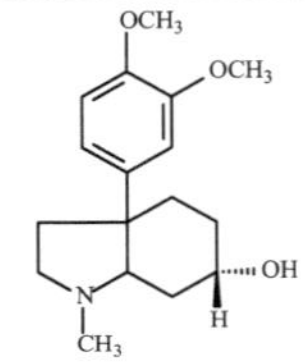	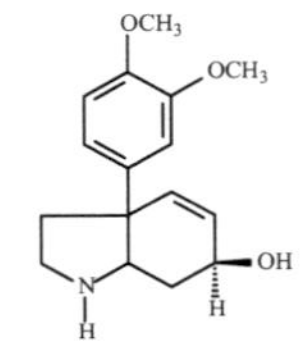
Alkaloid	(–)-Mesembrine	(–)-Mesembrine HCl	Mesembrenone	Δ^7 Mesembrenone	Mesembrenol	(–)-Mesembranol	Epimesembranol	(–)-N-Demethylmesembranol
MW	289.36	325.80	287.36	287.36		291.39	291.39	275.15
MF	$C_{17}H_{23}NO_3$	$C_{17}H_{23}NO_3$.HCl	$C_{17}H_{21}NO_3$	$C_{17}H_{21}NO_3$	$C_{17}\ H_{23}\ N\ O_3$	$C_{17}H_{25}NO_3$	$C_{17}H_{25}NO_3$	$C_{16}H_{23}NO_3$
Description	Pale yellow viscous liquid	Needle-shaped crystals	Pale yellow viscous liquid	Low melting solid	Pale brown crystalline powder	Cubic crystals	Pale brown oil	–
OR$[\alpha]_D^{20}$	-55.4° (MeOH)	-8.4° (MeOH)	racemic	–	–	-32°($CHCl_3$), -30° (C_2H_5OH)†	-3.2° (C_2H_5OH)†	-13°
BP	*186–190°C	–		–				
MP		205–206°C	†88–89°C	–		144–145°C		178–185°C
Reference	[4], *[13]	[4]	[4], †[7]	[4]		[4], †[7]	[4], †[7]	[4]

MW, Molecular weight; MF, molecular formula; OR, optical rotation; BP, boiling point; MP, melting point; MeOH, methanol.

Table 9. Physicochemical characteristics of mesembrine-type *Sceletium* alkaloids.

http://dx.doi.org/10.5772/66482

Structure	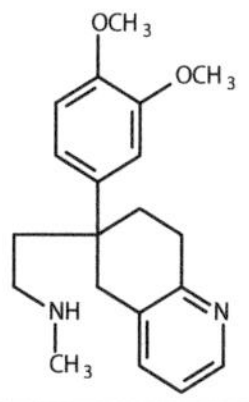	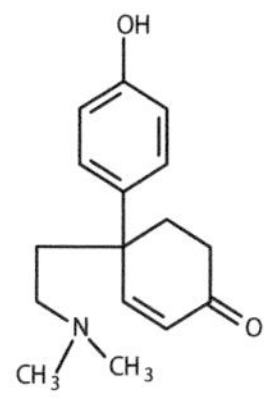	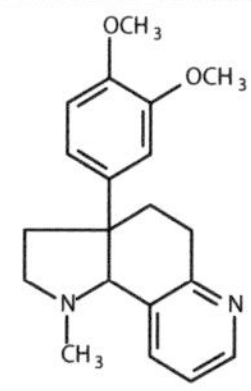
Alkaloid	Tortuosamine	Joubertiamine	*Sceletium* A4
MW	326	325.80	324.18
MF	$C_{20}H_{26}N_2O_2$	$C_{17}H_{25}NO_3$	$C_{20}H_{24}N_2O_2$
Description	–		Pale white semi-solid
OR$[\alpha]_D^{20}$		-32°($CHCl_3$)*, -30° (C_2H_5OH)†	*+131°
BP			
MP			†153–154°C
References	[4]	*[4], †[7]	*[4], †[8]

MW, Molecular weight; MF, molecular formula; OR, optical rotation; BP, boiling point; MP, melting point; MeOH, methanol.

Table 10. Physicochemical characteristics of some typical non-mesembrine-type *Sceletium* alkaloids.

The physicochemical characteristics of various Sceletium alkaloids—mesembrine-type and non-mesembrine-type alkaloids are compiled in **Tables 9** and **10**.

4. Extraction, isolation, synthesis and characterization of *Sceletium* alkaloids

Natural products are known to contain complex chemical components. Hence, it is essential that active components in such products are identified and analyzed by validated methods to ensure product quality. The development and validation of the requisite analytical method and procedures for QC can only be achieved by testing the product using qualified reference substances.

Several methods have been reported for the extraction and isolation of these alkaloids from *Sceletium* species. In 1937, Rimington and Roets [14] described their extraction procedure of *Sceletium* alkaloids, and subsequently in 1957, Bodendorf and Krieger [6] published a different extraction procedures. Popelak and Lettenbauer, in 1967 [7], reported the isolation of some alkaloid bases along with mesembrine and mesembrinine and prepared their hydrochloride salts. Arndt and Kruger [9] reported an extraction procedure of the aerial parts of *Sceletium joubertii* to obtain those relevant alkaloids.

Herbert and Kattah [10] in their biosynthesis study of alkaloids in *Sceletium subvelutinum* reported the isolation and purification of joubertiamine and related alkaloids. Jeffs et al. [8]

reported the extraction of alkaloids from *Sceletium namaquense* which yielded mesembrine, mesembrenone, *Sceletium* A4, N-formyltortuosamine, Δ^7mesembrenone, tortuosamine and some unidentified alkaloids. Smith et al. [15] extracted mesembrenol (Table 2, No. 9) {incorrectly designated as 4′-*O*-demethylmesembrenol and labeled (1) in their paper}, mesembrine and mesembrenone from Sceletium plant material. Gericke et al. [16] in their US patent application described the extraction of mesembrine-type alkaloids with a yield of between 15 and 35 mg per gram of "dry leaves."

Subsequently, Patnala [17] developed a relatively simple and inexpensive extraction and isolation procedure for *Sceletium* alkaloids. In general, Sceletium plant powder was extracted using ethanol by soxhlet extraction followed by alcohol removal and acidification. Hexane was used to wash the acidic solution and the organic phase discarded. Subsequently, ammonia solution was used to neutralize and result in alkaline solution, and the latter was further extracted with dichloromethane (DCM). The DCM fractions were collected into a round-bottomed flask and evaporated under vacuum to yield a brown viscous liquid containing alkaloids. Following the separation of components by column chromatography, collected eluents were spotted on a TLC plate (**Figure 4**). The TLC plate was first observed under UV_{254} which showed extensive related substances (*acetone-Track 3 and acetonitrile-Track* 4) and further sprayed with Dragendorff's reagent (**Figure 4**). The acetone fraction and the acetonitrile (ACN) fractions were found to contain alkaloids.

The ACN fraction was tested for its UV spectrum which showed a maximum at 298.2 nm was found to be Δ^7mesembrenone (**Figure 5**), and this fraction was further purified by preparative TLC.

In view of the fact that Sceletium species contain complex mixtures of closely related alkaloidal components, appropriate analytical methods for their separation and identification are

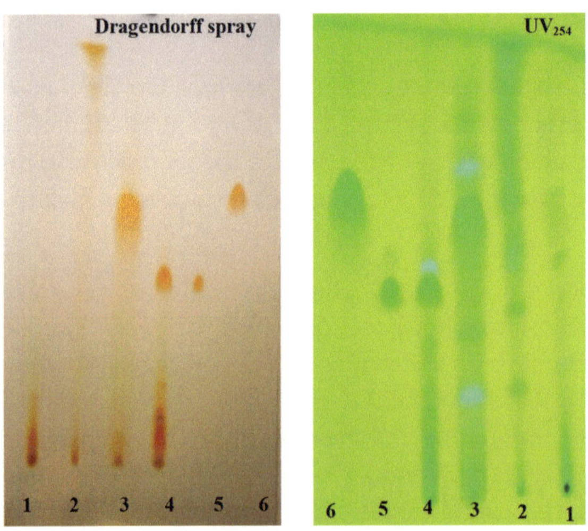

1. DCM fraction
2. Acetone-1 fraction
3. Acetone-2 fraction
4. ACN fraction
5. ACN fraction purified
6. Acetone-2 purified

Figure 4. TLC plate of the column fractions by developed TLC method observed under UV_{254} and subsequently sprayed with Dragendorff's reagent for positive identification of alkaloids.

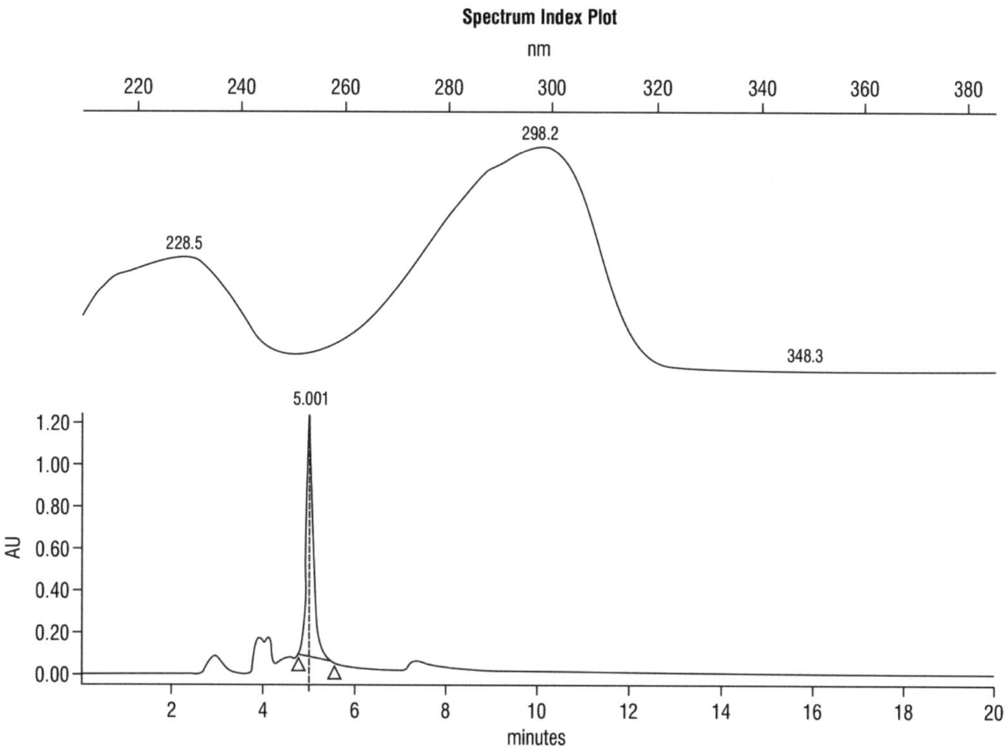

Figure 5. HPLC-PDA of ACN fraction-spectrum index plot (top) and chromatogram (bottom).

essential prerequisites for chemotaxonomic profiling of these species. Furthermore, the availability of relevant alkaloid reference standards is also necessary including the use of an analytical method with required specificity for fingerprinting. These foregoing considerations are essential to facilitate the proper identification of *Sceletium* species based on a chemotaxonomic approach [11].

5. Development of analytical methodologies for identification and quality control (QC) of *Sceletium* plant material and associated products

5.1. High-performance liquid chromatography (HPLC)

Chromatographic fingerprinting has been widely accepted and recommended by various regulatory authorities such as WHO [18], US-FDA [19] and EMEA [20] to assess the consistency of batch to batch dosage forms containing phytochemical components of the harvested plants. In the current international regulatory scenario, qualitative and quantitative analytical methods are considered mandatory.

Validated analytical methods to assay *Sceletium* plant material and dosage forms for relevant alkaloidal content were reported for the first time where a simple, accurate, precise, rapid and

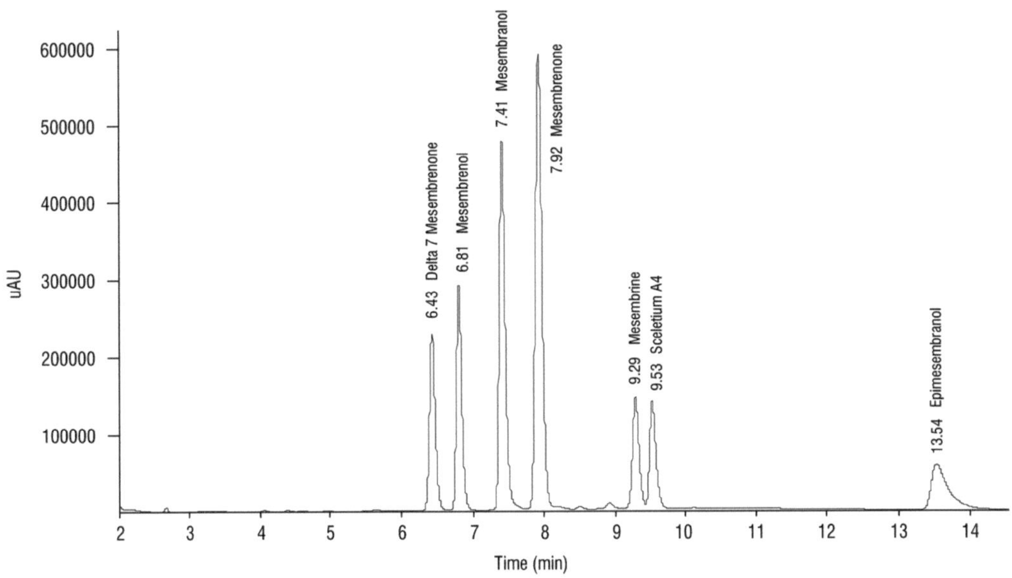

Figure 6. HPLC chromatogram of relevant standard Sceletium alkaloids.

reproducible HPLC method was developed for the identification and quantitative analysis of five relevant *Sceletium* alkaloids, Δ^7 mesembrenone, mesembranol, mesembrenone, mesembrine and epimesembranol. This method has also been successfully used to study chemotaxonomy of some *Sceletium* species and has provided impetus for the future development of quality monographs for plant and dosage forms containing *Sceletium* [21]. Subsequently, this method has been applied for the identification [22] and quantization of two additional alkaloids: Sceletium A4 and mesembrenol. **Figure 6** illustrates the chromatographic profile of the above-mentioned alkaloids.

5.2. LC-MS/MS

Since a number of variables including species differences, harvesting time, growing conditions, storage and processing contribute to the variation in phytochemical components in plants, it is therefore necessary to use appropriate and specific analytical methods to ensure quality which may affect the safety and efficacy of products prepared from plant material [23]. In particular, with respect to the *Sceletium* plant species which contain closely related mesembrine-type compounds of which some are epimers and have isobaric chemistries [17], specific methods are necessary. Since HPLC using UV detection cannot discriminate between such compounds, detection by MS enhances the accuracy and specificity of the analytical method, thereby reducing the risk of using an inappropriate *Sceletium* species for the indications on the product label [22]. In addition, this method proved valuable to monitor the fermentation process of *Sceletium* plant material [24]. Hence, the current qualitative LCMS method, and concurrent application of the previously reported quantitative assay method [22], provides valuable analytical procedures for the identification and QC of *Sceletium* plant material and its dosage forms. The application of the LC-ESI-MS tandem mass spectroscopy provides unique

http://dx.doi.org/10.5772/66482

fragmentation patterns which facilitates the identity of specific alkaloid in complex matrices and thus provides valuable confirmatory data for chemotaxonomic studies [11].

5.3. Capillary zone electrophoresis (CZE)

Since alkaloids are relatively strong bases in general [25], they are good candidates for CE analysis. A CZE method was developed and validated and applied to fingerprint the presence of alkaloids in a marketed tablet product containing Sceletium plant material [26].

6. Ethnopharmacology

The use of specific herbal medicines varies depending on specific regions and ethnopharmacological experiences, which makes this form of treatment inconsistent. Safety and efficacy are major concerns due to poor documentation and a dearth of scientific research on this subject. The World Health Organization (WHO) notes that of 119 plant-derived pharmaceutical medicines, about 74% are used in modern medicine in ways that correlate directly with their traditional use as herbal medicines [27].

The traditional preparation of *Sceletium* known as "*Kougoed*" or "*Channa*" is a fermented preparation used by the native Bushmen of Namaqualand. Traditionally, its main use for its psychoactive properties involved a prior fermentation by the Khoisan tribe of southern Africa, who purported that the psychoactive effect of this plant is greatly enhanced [2, 3]. Based on this perception, *Sceletium* plants and their products are marketed with claimed improvements in mood and reduction of anxiety, when the fermented plant material is used either by chewing or smoking. In general, the fermentation process involves crushing the whole plant material or aerial parts which are then placed in sealed containers for several days and dried under natural sunlight. Patnala [17, 24] subsequently confirmed that the fermentation process transforms mesembrine to Δ^7mesembrenone and requires an aqueous environment together with the presence of light to facilitate such a transformation.

7. Biological activities and medicinal properties of *Sceletium alkaloids*

The study of the phytochemical composition of Sceletium was provoked as a result of anecdotal information describing the use of these plants by early inhabitants of Southern Africa [28]. Typical examples of medicinal use have been described in the Ethnopharmacology section above. It can be gleaned from current scientific literature that several scientific groups working on various aspects of Sceletium plants have focused on the biological activity of these alkaloids [29]. It should be noted that antidepressant activity of mesembrine-type alkaloids has been demonstrated in animal models, of which, where mesembrine has been the principal alkaloid. The antidepressant activity is reportedly based on selective inhibition of serotonin reuptake, and mesembrine has a weak narcotic effect [30]. A recent study indicates that high-mesembrine Sceletium extract is a monoamine releasing agent, rather than only a selective serotonin reuptake inhibitor [31]. Zembrin® a marketed product containing a standardized extract of *Sceletium tortuosum* has been studied using human volunteers for its acute effects in the brain, and its pharmacological activity and potential therapeutic effect are reported to be based on the inhibiting reuptake of 5-HT and PDE4. It is suggested that a 25 mg dose of Zembrin® has the potential of reducing anxiety in humans [32].

8. Conclusions

Although eight *Sceletium* plant species have been formally classified in accordance with usual botanic taxonomy, we have observed the existence of various subspecies related to the tortuosum-type plants. Furthermore, the identified alkaloidal constituents vary between each of these plant species. In the tortuosum-type plants, mesembrenone (**Table 2**, No. 8), where the double bond occurs between C4-5; Δ7mesembrenone (**Table 3**, No. 15), where the double bond occurs between C7-7a; and the epimers, mesembranol and epimesembranol, clearly have closely related chemical structures. Hence, accurate identification and characterization is necessary to confirm the true identity of each species [22] in view of the close similarity between such chemical structures. such relevant information provides invaluable data to confirm the true identity of each species. These "tortuosum"-type Sceletium species contain mesembrine as the major alkaloid along with other minor alkaloids, Δ^7mesembrenone, mesembrenone and mesembranol and clearly differ from the other species. However, a subspecies of tortuosum type, *S. strictum* contains mesembrenone as the major alkaloidal component alongside mesembrine [11]. The above-mentioned information can be gleaned from published studies on *Sceletium* plants [17, 22, 24].

The advent and availability of modern instrumental techniques have provided valuable tools to identify differences between species based on phytochemical composition. Such approaches for taxonomical classification of plants and their species facilitate a superior and more accurate method which supersedes the classical techniques based on morphological aspects.

Although plants have been used for their medicinal properties for centuries relating back to biblical times, the interest and development of medicinal products containing plant material have grown exponentially where such products, often referred to a complementary medicines currently constitute and industry with sales of billions of dollars annually. However, there is growing concern relating to quality, safety and efficacy of such products where regulatory requirements relating to the provision of such necessary evidence currently leaves a lot to be desired and in instances have demonstrated undesirable risks to vulnerable users. Proper quality control requires the application of appropriate analytical techniques to assess the identity and quality of complementary medicines containing plant material. Quality control methods require access and availability to reference standards for each product which is marketed for medicinal use. As far as Sceletium-based products are concerned, the information relating to isolation, identification, quantification and purification of individual alkaloidal compounds found in Sceletium species provides valuable data for use in the quality control of medicines containing Sceletium plant material. While quality control is an essential component to ensure the quality of medicines, evidence of the safety and efficacy is further essential components, and it is important that such data are generated through clinical trials in humans. Furthermore, the absorption, distribution, metabolism and elimination (ADME) of administered products and associated kinetics should be studied. Such studies require the development and validation of appropriate analytical techniques to monitor the active ingredient(s) and the resulting metabolite(s) where applicable.

Modern instrumental methods such HPLC, LC-MS, CZE and associated analytical technologies have been invaluable in developing profiles for fingerprinting, identification and characterization of the relevant alkaloids and their specific plant associations as well as serving as an important tool for QC purposes of plant material and herbal medicines containing *Sceletium*. In addition, such techniques are also necessary to study the safety, efficacy, ADME and kinetics of medicinal products containing plant material.

Author details

Srinivas Patnala[1] and Isadore Kanfer[2, 3]*

*Address all correspondence to: izzy.kanfer@gmail.com

1 Basic Sciences Research Centre, KLE University, Belgaum, India

2 Faculty of Pharmacy, Rhodes University, Grahamstown, South Africa

3 Leslie Dan Faculty of Pharmacy, University of Toronto, Toronto, ON, Canada

References

[1] Gerbaulet M. Revision of the genus Sceletium N.E.Br (Aizoaceae). Botanishce Jarhbücher 1996;**118(1)**:9-24.

[2] Smith MT, Crouch NR, Gericke N, Hirst M. Phychoactive constituents of the genus Sceletium N.E.Br. and other Mesembryanthemaceae: a review. Journal of Ethanopharmacology 1996;**50**:119-130.

[3] Smith GF, Chesselet P, van Jaarsveld EJ, Hartmann H, Hammer S, Van Wyk B-E. Sceletium. Mesembs of the World. Pretoria: Briza Publications; 1998, p. 52.

[4] Jeffs PW. Sceletium Alkaloids. In: Manske RHF, Rodrigo RGA (editors). The Alkaloids. New York: Academic Press, Inc; 1981, p. 1-80.

[5] Rimington C, Roets GCS. Notes upon the isolation of the alkaloidal constituent of the drug "Channa" or "Kougoed" (Mesebryanthemum anatomicum and M. tortuosum). Ondersteproot Journal of Veternary Science and Animal Industry 1937;**9**:187-191.

[6] Bodendorf K, Krieger W. On the alkaloids of Mesembryanthemum tortuosum L. Archives Pharmacy 1957:**290/62(10)**:441-448.

[7] Popelak A, Lattenbauer G. The Mesembrine Alkaloids. In: Manske RHF (editor). The Alkaloids. New York: Academic Press; 1967, p. 467-481.

[8] Jeffs PW, Capps T, Johnson DB, Karle JM, Martin NH, Rauckman B. Sceletium alkaloids. VI. Minor alkaloids of S. namaquense and S. strictum. Journal of Organic Chemistry 1974;**39(18)**:2703-2710.

[9] Arndt RR, Kruger PEJ. Alkaloids from Sceletium joubertii.bol. The structure of joubertiamine and dehydrojoubertiamine. Tetrahedron Letters 1970;**37**:3237-3240.

[10] Herbert RB, Kattah AE. The biosynthesis of Sceletium alkaloids in Sceletium subvelutinum L. Bolus. Tetrahedron Letters 1990;**46(20)**:7105-7118.

[11] Patnala S, Kanfer I. Chemotaxonomic studies of mesembrine-type alkaloids in Sceletium plant species. South African Journal of Science 2013;**109(3/4)**:882-886.

[12] Martin NH, Rosenthal D, Jeffs PW. Mass spectra of Sceltium alkaloids. Organic Mass Spectrometry 1976;**11**:1-19.

[13] Mesembrine. Merck Index. 13th edn. New Jersey: Merck Research Laboratories; 2001.

[14] Rimington C, Roets GCS. Notes upon the isolation of the alkaloidal constituent of the drug "Channa" or "Kougoed" (Mesebryanthemum anatomicum and M. tortuosum). Ondersteproot Journal of Veterinary Science and Animal Industry 1937;**9**:187-191.

[15] Smith MT, Field CR, Crouch NR, Hirst M. The distribution of mesembrine alkaloids in selected TAXA of the mesembryanthemaceae and their modification in the sceletium derived 'Kougoed'. Pharmaceutical Biology 1998;**36(3)**:173-179.

[16] Gericke NP, Van Wyk B-E. Pharmaceutical Compositions Containing Mesembrine and Related Compounds. In US Patent: Gericke NP and Van Wyk B-E. 09/194,836 [US6,288,104B1]. 9-11-2006. South Africa; 1997.

[17] Patnala S. Pharmaceutical Analysis and Quality of Complementary Medicines: Sceletium and Associated Products (Thesis). Grahamstown: Rhodes University; 2007.

[18] General Guidelines for Methodologies on Research and Evaluation of Traditional Medicine. WHO/EDM/TRM/2000.1, Geneva: WHO; 2006; Annex1:21-26. Available from: http://apps.who.int/iris/bitstream/10665/66783/1/WHO_EDM_TRM_2000.1.pdf [Accessed: 2016-11-05].

[19] Guidelines for Dietary Supplements and Botanicals AOAC Official Methods of Analysis (2013) Appendix K, p. 2. Available from: http://www.eoma.aoac.org/app_k.pdf [Accessed: 2016-08-29].

[20] Guideline on Specifications: Test Procedures and Acceptance Criteria for Herbal Substances, Herbal Preparations and Herbal Medicinal Products/Traditional Herbal Medicinal Products. 2006: EMEA/CVMP/815/00 Rev 1. European Medicines Agency (EMEA). Available from: http://www.ema.europa.eu/docs/en_GB/document_library/Scientific_guideline/2009/09/WC500003393.pdf [Accessed: 2016-08-29].

[21] Patnala S, Kanfer I. HPLC analysis of Mesembrine-type alkaloids in Sceletium plant material used as an African traditional medicine. Journal of Pharmacy & Pharmaceutical Sciences 2010;**13(4)**:558-570.

[22] Patnala S, Kanfer I. Medicinal use of Sceletium: Characterization of phytochemical components of Sceletium plant species using HPLC with UV and electrospray ionization—tandem mass spectroscopy. Journal of Pharmacy & Pharmaceutical Sciences 2015;**18(4)**:414-423.

[23] Mauri P, Pietta P. Electrospray characterization of selected medicinal plant extracts. Journal of Pharmaceutical and Biomedical Analysis 2000;**23(1)**:61-68.

[24] Patnala S, Kanfer I. Investigation of phytochemical content of Sceletium tortuosum following the preparation of "Kougoed" by fermentation of plant material. Journal of Ethnopharmacology 2009;**121**:86-91.

[25] Unger M. Capillary zone electrophoresis of alkaloids Influence of structure on electrophoretic mobility. Journal of Chromatography A 1998;**807**:81-87.

[26] Patnala S, Kanfer I. A capillary zone electrophoresis method for the assay and quality control of mesembrine in Sceletium tablets. Journal of Pharmaceutical and Biomedical Analysis 2008;**48**:440-446.

[27] Cunningham AB. An Africa-Wide Overview of Medicinal Plant Harvesting, Conservation and Health Care. In: Bodeker G, Bhat KKS, Burley J, Vantomme P (editors). Medicinal Plants for Forest Conservation and Health Care. Food and Agriculture Organization of the United Nations. p. 123. Available from http://www.fao.org/3/a-w7261e.pdf [Accessed 2016-10-14].

[28] Pappe L. An Enumeration of South African Indigenous Plants Used as Remedies by the Colonists of the Cape of Good Hope. In: Florae Capensis Medicae Prodromus. 3rd edn. Cape Town, South Africa: W Brittain; 1868. http://kanna-info.com/history-of-kanna Accessed [2016-12-05]

[29] Carpenter JM, Jourdan MK, Fountain EM, Ali Z, Abe N, Khan IA, Sufka KJ. The effects of Sceletium tortuosum (L.) N.E.Br. extract fraction in the chick anxiety-depression model. Journal of Ethnopharmacology 2016;**193**:329-332.

[30] Smith C. The effects of Sceletium tortuosum in an in vivo model of psychological stress. Journal of Ethnopharmacology 2011;**133**:31-36.

[31] Coetzee DD, Lopez V, Smith C. High-mesembrine Sceletium extract (Trimesemine™) is a monoamine releasing agent, rather than only a selective serotonin reuptake inhibitor. Journal of Ethnopharmacology 2016;**177**:111-116.

[32] Terburg DD, Syal S, Rosenberger LA, Heany S, Phillips N, Gericke N, Stein DJ, Honk JV. Acute effects of Sceletium tortuosum (Zembrin®), a dual 5-HT reuptake and PDE4 inhibitor, in the human Amygdala and its connection to the hypothalamus. Neuropsychpharmacology 2013;**38**:2708-2716.

Chapter 5

Reducing of Alkaloid Contents During the Process of Lactic Acid Silaging

Annett Gefrom and Annette Zeyner

Additional information is available at the end of the chapter

http://dx.doi.org/10.5772/66287

Abstract

The current study is conducted to investigate whether lupine grains can successfully be ensiled with residual moisture contents of around 65% and alkaloid contents can be reduced during the process of lactic acid formation by native and added lactic acid bacteria, respectively. Based on the fermentation quality of the sampled grain meal silage in the dry mass of 65% and further results, silaging is shown to be a suitable method of preservation. A reduction in the alkaloid content during the silaging cannot be assumed (statistical) due to the irregular dynamic of the observed content.

Keywords: lupines, lactic acid moist grain silaging, nutritional facts and alkaloid content, lupine debittering

1. Introduction

At present, soy is the most important protein feed for monogastric farm animals and also of high importance for polygastric species, particularly in face of the ban of feedstuffs of animal origin. To cover the high demand for animal nutrition, the European Union is dependent on the import of soybeans and soy products. The European agricultural policy is focused on alternative protein sources because of the ongoing debate on the import of genetically modified soy. Additional arguments for maximum self-supply with protein from local sources are the goals of sustainable production, the fluctuation of prices in soy bean business, and the future competition between food and energy production.

Legume grains like those from lupine species are valuable feedstuffs in this concern. Beside the remarkable protein content, they impress by a high density of energy.

Due to the uneven ripeness of the stock (**Figure 1**) and particularly if harvested at moist conditions, for example, in late summertime, a cost-intensive artificial drying of the grains is necessary. Storing lupine grains with a residual moisture above 14% might increase the risk for elevated counts of moulds and yeasts.

Figure 1. Uneven ripeness of lupine corns (picture: A. Gefrom, LFA MV).

The chemical conservation with soda lye, propionic acid or feed urea is possible at a moisture of <20–25%, but demands a Hazard Analysis and Critical Control Concept. The material may be storable up to 6 months but might be limited in their use with respect to the target animal species or specific situation in question. Furthermore, chemical conservation of feedstuffs is not permitted for organic farming.

Ensiling lupine grains may be an inexpensive and ecologically advantageous alternative to produce a high-protein feed of local origin that can be used in both conventional and organic farming. Lactic acid fermentation of fully mature legume grains with accordingly high dry matter contents is, however, impossible because the osmotolerance of the vast majority of lactobacilli is overused under such conditions. To create good conditions for silaging, lupine grains can be harvested either fully mature and remoistened or before full ripeness with accordingly high moisture content. According to technical aspects, harvesting before full ripeness and conservation by lactobacilli have the following particular benefits:

- Preservation of the feed can be performed by the farmer himself directly after harvest.
- Independence of harvest date and dry matter content (possible with moisture contents of up to about 35%).
- Early field clearance and therefore effective use of the land.

- Minimisation of nutrient loss (storage and threshing loss).
- Reduced costs through elimination of the artificial drying.
- Cost-effective production of feedstuff with high energy and protein content.
- Providing high-quality feedstuffs rich in protein and energy for farm animals in both conventional and organic farming systems.

The current study was conducted to investigate whether lupine grains can successfully be ensiled with residual moisture contents of around 65%, and alkaloid contents can be reduced during the process of lactic acid formation by native and added lactic acid bacteria, respectively. Hypotheses were that high-quality silages with reduced alkaloid contents can be produced in this way.

Alkaloids are important anti-nutritional factors limiting the use of lupine grains in the nutrition of farm animals. Due to the very effective alkaloids, low dose could induce physiological effects. Alkaloids damage among others the central nervous system, respiratory tract, and liver [1, 2]. Because of that, native bitter lupine grains are not suitable as feedstuff in particular for monogastric animals. For sweet lupine varieties provided as feed, actual guidelines demand a maximum alkaloid content of 0.05% to exclude intoxication [3, 4]. Further, the bitter taste can lead to a reduced acceptance and feed intake thus causing decreased performance. Pigs are particularly sensitive in this concern. However, under the environmental conditions of high temperatures, the accumulation of alkaloids may vary, and the alkaloid content in sweet lupines may be higher than the alkaloid limit for using lupines as feedstuff [5].

On the other hand, alkaloids are part of the protective system of the plant against pathogens. Alkaloids are nitrogenous combinations and attributive to some of the most toxic plant ingredients. Because of the molecular structure, the alkaloids in lupines allocate to chinolizidine alkaloide [6–8]. In corns of domestic lupines appear predominantly main quinolizidine alkaloids like spartein, lupinin, and lupanin in aggregate relevant to the lupine genre [9]. Other alkaloid contents in lupines are under 1%.

White lupines	Lupanin (50–80%), multiflorin (3–10%), 13-hydroxylupanin (5–15%), albin (5–15%)
Blue lupines	Lupanin (50–80%), angustifolin (5–20%), 13-hydroxylupanin (10–20%)
Yellow lupines	Lupinin (40–70%), spartein (30–50%)

The alkaloids differ in toxicity. Spartein has a lower LD50 (median lethal dose) as lupanin [10]. Chinolizidine alkaloids are located are located in all plant organs, but the concentration and variation differ in texture, seasonal and daytime [9, 11, 12]. They are accumulated in corns and converted to proteins during maturation. They were also used as nitrogen source during germination [13–15].

It is possible that especially at not determinated varieties, green corns of the bastard branch have higher alkaloid content. Under the environmental conditions of high temperatures, the accumulation of alkaloids may vary, and the alkaloid content in sweet lupines may be higher than the alkaloid limit for using lupines as feedstuff [5, 16–18]. A post-harvesting reduction in alkaloids through silaging may allow future cropping decisions basing more on phytosanitary conditions. The aim of this work was to examine whether alkaloids can be reduced in their concentration by fermentation with lactobacilli on the precondition that preservation process per se succeeds.

Alkaloids can be reduced by hydrolytic and oxidative enzymes during germination [13]. From a theoretical point of view, alkaloids may be decomposed throughout silaging under the action of native and supplemented lactic acid bacteria, respectively. A comprehensive action of plant and microbacterial enzymes is possible due to the grinding of the grains, the setting of suitable osmotic conditions, and a sufficiently long fermentation time.

The moisture found in grain meal allows enzymes to process the substances in suspension freed in the fluid phase of fermentation. Phytonutrients exist in the wet phase in soluble form and change through plant enzymes and enzymes of microorganisms occurring during fermentation [19, 20], so that phytonutrients can turn to low molecular weight units and into milk acid [21].

It has been demonstrated that the alkaloid content of lupine grains in milk ripeness can be reduced through silaging [22]. On the contrary, only small success in this concern was reported following silaging of the whole bitter lupine plant where the alkaloids seemed to be fairly stable in the acid milieu of the silage. It is assumed that alkaloids inhibited lactic acid bacteria during the silaging process [23].

Hopper et al. [24, 25] and Santana et al. [20], however, described a possible elimination of bitterness in lupines, since alkaloids serve as energy source for microorganisms. In this way, Toczko [26] and Toczko et al. [27] demonstrated the conversion of lupanin through Pseudomonas sp. These bacteria are representatives for natural epiphytics which are proposed to be particularly important in the first phase of the process of fermentation. Santana and Empis [28] reduced the alkaloid content in flour of white bitter lupine seeds by 50% by incubation with Pseudomonas for 4 days. Camacho et al. [29] also demonstrated a reduction in alkaloids in lupine grains by incubation with *Lactobacillus acidophilus* (B-1910) in a 12% total solids suspension.

Apart from the possible reduction in alkaloids during fermentation, further benefits like physiologically effects of lactic acid silages of moist lupine grains are as follows:

- The positive effects of lactic acid in terms of nutrition physiology, for example, the acidification of distinguished parts of the digestive tract to prevent the proliferation of clostridia and other pathogenic microorganism.

- The improvement in the feeding value due to the reduction in other anti-nutritional factors such as oligosaccharides [30, 31].

- Elevated contents of essential amino acids following proteolysis but only when desmolysis can be prevented [32].

- Improved intake and digestion of the feed [33].

http://dx.doi.org/10.5772/66287

- Improved digestibility of the amino acids (demonstrated for lysine and methionine by Hackl et al. [33]).
- Enabling to elevate the quantity of lupine grains in the daily ration (broiler, 10%; weaned piglets, 12%) and thus, elevated waiver of soy products [34].

2. Reducing of alkaloid contents during the process of lactic acid silaging

2.1. Material and method

The experiments were performed with ripe, storage-dry lupine grains (variety of sweet lupines 'Bora' and bitter lupine 'Azuro'; 2002) as well as with legume grains from different years (2005 and 2006) with a high residual moisture content (~65% DM).

In crushed seeds (sieve 3 mm; **Figure 2**) of ripe, storage-dry lupine grains and at ~65% DM harvested *Lupinus angustifolius* L. (variety of sweet lupines 'Bora', 'Borlu', and bitter lupine 'Azuro') were determined the nutrient contents according to the Weende feed analysis [35]. Crude starch and water-soluble carbohydrates (fructose, glucose, sucrose, and galactose) were analysed by HPLC (Shimadzu-Germany GmbH) [36].

Figure 2. Crushed lupine seeds (picture: A. Gefrom, LFA MV).

The material was ground by ball mill (swing mill 'MM 200' Retsch GmbH & Co. KG, 42781 Germany, grain size 0.01 mm, 5 min with frequency of 30/s) for analysis of starch and alkaloids. The total alkaloid content (quinolizidine alkaloids) was determined by GC-MS [37]:

Preparation	To weigh ca. 0.5 g sample in centrifuge tubes, add 20 ml 0.5 N salt acid homogenize and store for 16 h at room temperatures; centrifugalize by 4000 U/min for 30 min and by 20°C; pipette overlap (determine volume); add 6 N NaOH for adjust pH-Wert of 12 in overlap (mix); alkaloids were determinated with solid phase extraction with extrelut column (Merck) and dichlormethan as eluent: add 18 ml to extrelut column (determine volume); wait 15 min; elute with 3 × 20 ml dichlormethan; solvent was evaporated and rest is absorbed with 1 ml methanol; sample could be stored by −20°C
Analysis	Alkaloids were analysed by GC-MS (Shimadzu-Germany GmbH, GC-MS QP 2010), type of column: BP-1 (length: 30 m; diameter: 0.25 mm; thickness: 0.25 μm); split proportion: 50; carrier gas: helium; passage: 94 ml/min; start temperature:120°C; end temperature: 260°C; ion source: 230°C; interface temperature: 300°C; externer standard: lupanin and spartein

The preparation of model silage (ROMOS—Rostock model silages [38]; **Figure 3a** and **b**) with 65% DM (moistened ripe, storage-dry lupines meal and ~65% DM harvested lupine corns) followed using various biological silage additives (600 g material per plastic bag (three repetitions per variation)):

Figure 3. (a and b) Vacuum sealer and ROMOS—Rostock model silages (picture: A. Gefrom, LFA MV).

- Control (without additive)
- Lactic acid bacteria (LAB, homofermentative, *Lactobacillus plantarum*; 3 × 10^5 cfu/g FM; DSM 8862, 8866)

By using a vacuum sealer, the bags got evacuated and sealed and afterwards stored at 20°C for 90 days (**Figure 4**). The incubation was followed by the pH-measurement (WTW MultiCal pH 526, precision: 0.01), the analysis of ammonia nitrogen and the analysis of organic acids and alcohol by HPLC and gas chromatograph (Shimadzu) in the silage extracts (50 g silage + 200 ml aqua dest.; 15 h at 5°C). The contents of lactic acid, fatty acid, and alcohol were analysed by HPLC. Gas chromatograph with standard was used for analysis of acetic acid, butyric acid, propionic acid, and alcohol (ethanol, propanol, butanol, butandiol). Parameter settings of HPLC and gas cromatograph are as follows:

HPLC:	Shimadzu-Germany GmbH; separation column: Aminex HPX-87H, Biorad, Hercules, USA; flux 10 mM H_2SO_4, flow rate: 0.6 ml/min, temperature: 60°C and detektion: UV index by 210 nm
Gas chromatograph: Shimadzu with integrator chromatpac (GC14A)	Dose 0.5 μl with split; capillary column: FFAP-DF-0.25; 25 m × 0.32 mm; carrier gas: N_2 purity; P1 = 0.75 kp/cm²; P2 = 1 kp/cm²; hydrogen = 0.6 kp/cm²; air = 0.5 kp/cm²; temperature program: 1.5 min by 110°C constant, heating phase: 12°C/min by 170°C; 3 min by 170°C constant; injector temperature: 190°C; detektion: FID (Shimadzu), temperature: 190°C; detector sensitivity: 10^2; split ratio: 1:50–1:70; inside standards: iso-capronic acid for volatile fatty acids, n-pentanol for alcohol

Figure 4. Model silages after 90 days (picture: A. Gefrom, LFA MV).

Evaluation of parameters occurs according to DLG [39].

In freeze-dried silage material nutritional parameters were analysed according to the Weende feed analysis and alkaloids were analysis by GC-MS [37].

For statistical analysis, the computer software SPSS 14 for Windows (SPSS, Chicago, IL, USA) was used. Duncan test was applied to examine mean differences for significance. The level of significance was preset at $p < 0.05$.

3. Results

3.1. Nutrient contents of lupine seeds and silages after incubation of 90 days

The high energy and protein contents of moist harvested lupine grains from this study are in accordance with those reported by key feed tables (**Tables 1** and **2**). According to Jansen et al. [40], White et al. [41], and Wrigley [42], the crude starch in lupine corns analysed with enzymatical method contains approximately 2.4% in DM and is lower than contents after analysis with polarimetric method used for declarations in DLG. These authors supposed that crude starch contents analysed with polarimetric method could include sugar and non-starch polysaccharides.

	DLG (2014)[1]	Starting material 'Bora', 'Borlu', 'Azuro'; 2005/06, *n* = 6	
Crude ash [% DM]	3.6	3.7	±0.2
Crude protein [% DM]	33.5	35.8	±0.9
UDP [% XP]	20	–	
nXP [%]	21.9	–	
RNB [g N/kg]	+19	–	
Crude fat [% DM]	5.5	6.2	±0.7
Crude fibre [% DM]	16.3	15.0	±1.8
Crude starch [% DM]	[2]6.0	[3]2.2	±0.2
Crude sugar [% DM]	5.6	4.0	±1.0
NEL [MJ/kg DM]	8.92	8.95	±0.09
ME_{cow} [MJ/kg DM]	14.19	14.27	±0.13
ME_{pig} [MJ/kg DM]	15.3	14.4	±0.19
$AME_{Npoultry}$[4] [MJ/kg DM]	8.8	8.02	±0.36

[1]DLG 2014 [43] nutrient digestibility ruminants and UDP (undegrable protein): DLG [44].
[2]Polarimetric method.
[3]Analysed by HPLC with enzymatic method; DM: dry matter; LAB: lactic acid bacteria [*Lactobacillus plantarum* 3×10^5 cfu/g fresh matter (FM), DSM 8862, 8866]; ME_{pig}: metabolizable energy for pigs; calculated according to the estimation equation from GfE [45] and digestibility from DLG [44].
[4]$AME_{N\ poultry}$: apparent metabolizable energy for poultry (N-corrected): calculated according WPSA [46]; NEL MJ/kg DM and MEcow MJ/kg DM: energy calculated according to DLG [44]; nXP: available crude protein calculated according to DLG [44]; RNB: ruminal N-bilance, RNB = (XP–nXP)/6.25; UDP: undegrable protein.

Table 1. Contents of proximate nutrients and energy in grains of blue lupine from literature (mature seeds, in 100% DM) compared to experimental data with lupine seeds harvested at ~65% DM (blue lupine).

Blue lupine 'Bora' 2005, n = 3			Silage			
			Control		LAB	
DM [%]	67.2^{a}	±0.1	66.2^{b}	±0.6	65.9^{b}	±0.2
Crude ash [% DM]	4.0^{a}	±0.1	3.9^{b}	±0.1	3.8^{c}	±0.1
Crude protein [% DM]	35.8^{b}	±0.3	36.9^{a}	±0.6	36.6^{ab}	±0.4
Crude fat [% DM]	5.8	±0.0	6.1	±0.3	6.1	±0.2
Crude fibre [% DM]	16.5^{a}	±0.1	15.8^{b}	±0.5	15.8^{b}	±0.2
Crude starch [% DM][1]	2.4^{a}	±0.2	0.7^{c}	±0.3	1.2^{b}	±0.6
WSC [% DM]	4.9^{a}	±0.6	2.3^{b}	±0.7	2.5^{b}	±0.6
NEL [MJ/kg DM]	8.89	±0.01	8.99	±0.02	8.99	±0.02
ME_{cow} [MJ/kg DM]	14.19	±0.01	14.32	±0.05	14.33	±0.04
ME_{pig} [MJ/kg DM]	14.24	±0.02	14.33	±0.08	14.37	±0.06
$AME_{Npoultry}$[2] [MJ/kg DM]	8.16	±0.03	8.08	±0.15	8.11	±0.09

[1]Analysed by HPLC; with enzymatic method; control: without additive; DM: dry matter; LAB: lactic acid bacteria [*Lactobacillus plantarum* 3 × 10^5 cfu/g fresh matter (FM), DSM 8862, 8866]; ME_{pig}: metabolizable energy for pigs; calculated according to the estimation equation from GfE [45] and digestibility from DLG [44].
[2]$AME_{N\ poultry}$: apparent metabolizable energy for poultry (N-corrected): calculated according WPSA [46]; NEL MJ/kg DM and MEcow MJ/kg DM: energy calculated according to DLG [44]; WSC: water soluble carbohydrates: fructose, glucose, sucrose, galactose
[a,b]Significant ($p < 0.05$) differences of means between variants starting material (lupine 'Bora') and silages.

Table 2. Contents of proximate nutrients and energy in grains of blue lupine harvested at ~65% DM and therefrom produced silages after incubation of 90 days.

The content of nutritional feed value parameters is, apart from the fermentable carbohydrates (WSC), not affected by the silaging process, and the high feed and energy content of legumes are maintained in the grain silage.

3.2. Quality of silages

To ensure optimal osmotic conditions for silaging, lupine grains can either be harvested as mature grains and remoistered immediately before being ensiled, or harvested immature and ensiled as harvested. Based on the fermentation quality of the sampled grain meal silage in the dry mass range of ~65% DM and further results, silaging is shown to be a suitable method of preservation (**Table 3**). Harvest-moist grains may be stored without problems, even without the addition of silage additives. The addition of high-performance lactobacilli ensures the fermentation through an earlier and more comprehensive production of lactic acid. The production of acetic acid and alcohol was reduced.

	DM		pH		LA		AA		ΣAL		NH_3-N	
	[%]				[% DM]						[% N]	
SM	66.5	±0.6	5.9^a	±0.1	n. a.							
CON	66.3	±0.6	4.8^c	±0.3	2.5^b	±1.0	0.5^b	±0.2	0.3^a	±0.2	1.0^a	±0.1
LAB	66.1	±0.4	4.2^d	±0.1	5.3^a	±0.4	0.6^a	±0.1	0.1^b	±0.0	0.7^b	±0.1

AA: acetic acid; CON: control (without additive); control without additive; DM: dry matter; LA: lactic acid; LAB: addition of lactic acid bacteria [*Lactobacillus plantarum* 3 × 10^5 cfu/g fresh matter (FM), DSM 8862, 8866]; n. a.: not analysed; NH_3-N: ammonia nitrogen; SM: starting material; ΣAL: alcohol (ethanol, propanol, butanol, butandiol). a, b Significant ($p < 0.05$) differences of means between variants.

Table 3. Fermentation parameters of silages from lupine grains (harvested at ~65% DM) 90 days of ensiling and ensiled with or without lactic acid bacteria (LAB) as additive (own data, means of all samples from 2005 ('Bora', 'Borlu', 'Azuro'; n = 18).

Variety						**DM**		**Alkaloids**	
					n	[%]		[% DM]	
Bitter lupine	Dry seed	'Azuro'	SM		1	65.0	±0.30	2.289	±1.09
			Silage	Control	6	63.5	±0.72	2.246	±0.34
				LAB	18	64.4	±0.85	2.615	±0.95
	Moist grains		SM		3	66.1	±0.1	2.991	±1.42
			Silage	Control	6	66.6	±0.2	2.765	±0.50
				LAB	6	66.1	±0.7	2.743	±0.22
Sweet lupine		'Borlu' + ''Bora'	SM		6	67.2	±0.1	0.109^a	±0.02
			Silage	Control	12	66.1	±0.7	0.079^b	±0.01
				LAB	12	66.1	±0.2	0.074^b	±0.02
		'Borlu'	SM		3	67.2	±0.0	0.116^a	±0.03
			Silage	Control	6	66.1	±0.9	0.082^b	±0.01
				LAB	6	66.2	±0.2	0.078^b	±0.01
		'Bora''	SM		3	67.1	±0.1	0.103^a	±0.02
			Silage	Control	6	66.2	±0.6	0.075^b	±0.01
				LAB	6	65.9	±0.2	0.069^b	±0.02

Control without additive; DM: dry matter; LAB: addition of lactic acid bacteria [*Lactobacillus plantarum* 3 × 10^5 cfu/g fresh matter (FM), DSM 8862, 8866]; SM: starting material; alkaloids: external standard at measurement with GC-MS: lupanin and spartein; a, b significant ($p < 0.05$) differences of means between variants.

Table 4. Alkaloids in dry corn from year 2002 ('Azuro') and in moist harvested grains (2005) from sweet lupines ('Borlu' and 'Bora') and bitter lupine ('Azuro') and their silages after 90 days incubation.

3.3. Alkaloids

Table 4 shows the alkaloid content of lupine seeds (bitter 'Azuro' and sweet variants 'Bora') harvested at usual dry matter (2002) and approximate 65% DM (2005) and their silages after incubation of 90 days. The alkaloid content in analysed corns of bitter lupines ('Azuro') is higher than sweet lupines. Bitter lupines can contain high alkaloid contents of 4.5% DM [47, 48]. According to studies at the Julius Kühn Institute, the variety of the bitter lupine 'Azuro' contains 1.375% alkaloids in DM [29]. Sujak et al. [49] published a variety of alkaloids in sweet lupines between 0.05 and 0.24% in DM. Jansen et al. [16] published alkaloid contents in blue sweet lupines between 0006 and 0068% in DM. But under the environmental conditions of high temperatures, the accumulation of alkaloids may vary, and the alkaloid content in sweet lupines may be higher than the alkaloid limit for using lupines as feedstuff [5, 16, 17]. In blue sweet lupines 'Borlu' the alkaloid content reached a high level of 0.116% in DM and illustrated the variability between years. Otherwise, the grains were harvested earlier than normal, and the crop could contain more green corns with maybe high alkaloid content.

Against the statement of Camacho et al. [29], a reduction in alkaloid content in lupine silages through the fermentation process could not initially be shown because of the low differences and irregular dynamic in alkaloid contents. In all likelihood, the alkaloids are stabile in an acid environment [23]. A change in toxic potential of alkaloids in lupines after silaging should be reviewed in experimental feeding tests.

Pearson and Carr [50], Godfrey et al. [51], Bellof and Sieghart [52], and Allen [53] defined an experimental limit of tolerance of alkaloid contents in ration for feeding monogastric animals of 0.03%. But also contents of 200 mg alkaloid/kg could reduce feeding intake and growth of the animals [50, 51]. Petersen and Schulz [54] analysed an alkaloid content of 0.01% that effected the feeding intake.

4. Conclusions

Making silages of moist legume grains could be a lower-cost and ecological conservation process for the production of protein-rich feed in conventional and organic farming.

From the results of the experiments with fermented grain meal with high moisture content, it is possible to draw the following conclusions:

1. Harvest-moist grains at 65% DM are possible.
2. Lupine grains are a valuable feed due to their high energy and protein content and their composition. There are no degradations of nutritional parameters when harvested at dough-ripe stage, because dimerization is finished.
3. Despite unfavourable chemical silage properties, the preservation of legume grains is possible even with a high amount of dry matter.

4. The preservative effect of acid formation is primarily dependent on the moisture content. In respect of the process safety in industrial conditions, moisture content of 35% for lactic acid silage inputs is to be recommended.

5. Lactic acid bacteria stabilize the ensiling process and the silage quality.

6. The content of nutritional feed value parameters is, apart from the fermentable carbohydrates, not affected by the silaging process, and the high feed and energy content of legumes are maintained in the grain meal silage.

7. A reduction in the alkaloid content during the silaging cannot be assumed (statistical) due to the irregular dynamic of the observed content. The number of samples should be considered for future investigations. Maybe other bacillus could reduce alkaloid contents.

   In own studies, the alkaloid contents could decrease tendencially by additional different lactobaccilus. A possible change in toxicity of the alkaloids during fermentation should be investigated by feeding studies.

8. The fermentation of early-harvested legume grains with high moisture content can therefore be recommended as a suitable method of conservation. With respect to the economic and organisational benefits when compared to the discussed methods of chemical conservation, the lactic acid fermentation of grains under anaerobic conditions conforms to the requirements of organic agriculture and is therefore also interesting for this branch. Silaging in plastic tubes is recommended (**Figure 5**).

Figure 5. Silaging in plastic tubes is recommended (picture: A. Priepke, LFA MV).

Acknowledgements

The study was financed by Bundesministerium für Ernährung und Landwirtschaft.

We are grateful to the H. Wilhelm Schaumann Stiftung for funding a scholarship.

Author details

Annett Gefrom[1*] and Annette Zeyner[2]

*Address all correspondence to: annett.gefrom@gmx.de

1 State Research Centre of Agriculture and Fishery Mecklenburg-Western Pomerania, Germany

2 Group Animal Nutrition, Faculty of Natural Sciences III, Institute of Agricultural and Nutritional Sciences, Martin Luther University Halle-Wittenberg, Halle (Saale), Germany

References

[1] Waldroup PW, Smith KJ. Animal feed uses of legumes. In: Matthews RH, editor. Legumes. Chemistry, Technology and Human Nutrition. Marcel Dekker Inc., New York und Basel, 1989. pp. 245-327. doi:19911433828

[2] Birk Y. Antinutritional factors (ANFs) in lupins and in other legume seeds. In: Proceedings of the 7th International Lupin Conference; 18-23 April 1993; Évora. Portugal. pp. 424-429.

[3] Garcia AG. Cultivos herbaceous extensivos. Mundi-Prensa, Madrid, 1984.

[4] Jeroch H. Körner und Samen. In: Jeroch H, Flachowsky G, Weissbach F, editors. Futtermittelkunde. 1st ed. Gustav Fischer Verlag Jena, Stuttgart, 1993. pp. 281-293. ISBN:9783334003848

[5] Jansen G, Jürgens HU, Ordon F. Effects of temperature on the alkaloid content of seeds of *Lupinus angustifolius* Cultivars. Journal of Agronomy and Crop Science. 2009;195:172-177. doi:10.1111/j.1439-037X.2008.00356.x

[6] Petterson M. Composition and food uses of lupins. In: Gladstones JS, Atkins CA, Hamblin J, editors. Lupins as crop plants: biology, production and utilization. Wallingford: CAB International, Cambridge; 1998. pp. 353-384. ISBN:0851992242

[7] ANZFA (Australia New Zealand Food Authority) (2001). Lupin alkaloids in food—a toxicological review and risk assessment. Technical Report, Series 3, Canberra; Available from: http://www.foodstandards.gov.au/publications/documents/TR3.pdf [Accessed: 2016-09-27].

[8] Wink M. Evolution of secondary metabolites from an ecological and molecular phylogenic perspective. Phytochemistry. 2003;64(1):3-19. doi:10.1016/S0031-9422(03)00300-5

[9] Wink M. Methoden zum Nachweis von Lupinen-Alkaloiden. In: Wink M, editors. Lupinen 1991 -Forschung, Anbau und Verwertung. Universität Heidelberg; 1992. 78-90.

Available from: http://www.uni-heidelberg.de/institute/fak14/ipmb/phazb/pubwink/1992/27.1992.pdf [Accessed: 2016-09-27].

[10] Wink M. Biological activities and potential application of lupin alkaloids. In: Proceedings of the 7th International Lupin Conference; 18-23 April 1993; Évora, Portugal, pp. 161-178.

[11] Wink M, Witte L. Turnover and transport of quinolizidine alkaloids: diurnal variation of lupanine in the phloem sap, leaves and fruits of *Lupinus albus* L. Planta. 1984;161:519-524.

[12] Lee MJ, Pate JS, Harris DJ, Atkins CA. Synthesis, transport and accumulation of quinolizidine alkaloids in *Lupinus albus* L. and *L. angustifolius* L. Journal of Experimental Botany. 2007;58(5):935-946. doi:10.1093/jxb/erl254

[13] Wink M. Stoffwechsel und Funktion von Chinolizidinalkaloiden in Pflanzen und pflanzlichen Zellkulturen [thesis]. University Braunschweig; 1984.

[14] Wink M. Chemische Verteidigung der Lupinen: Zur biologischen Bedeutung der Chinolizidinalkaloide. Plant Systematics and Evolution. 1985;150:65-81. doi:10.1007/BF00985568

[15] Wink M, Witte L: Quinolizidine alkaloids as nitrogen source for lupin seedlings and cell cultures. Zeitschrift für Naturforschung. 1985;40c:767-775.

[16] Jansen G, Jürgens HU, Beyer H, Seddig S. Alkaloidgehalt in Blauen, Gelben und Weißen Lupinen [Internet]. Available from: http://lupinenverein.de/wp-content/uploads/2014/04/Alkaloidgehalt-in-Lupinen.pdf [Accessed: 2016-09-02].

[17] Jansen G, Jürgens HU, Flamme W. Breeding of sweet lupines for the organic farming —quality investigations according to feed Suitability [Internet]. Available from: http://forschung.oekolandbau.de [Accessed: 2016-09-02].

[18] Christiansen JL, Jornsgard B, Buskov S, Olsen CE. Effect of drought stress on content and composition of seed alkaloids in narrow-leafed lupin, *Lupinus angustifolius* L. European Journal of Agronomy. 1997;7:307-314. ISSN:1161-0301

[19] McDonald P, Henderson AR, Heron SJE. The biochemistry of silage. 2nd ed. Cambrian Printers Ltd: Aberystwyth; 1991. 340 p. doi:19811424455

[20] Santana FMC, Fialho AM, Sa-Correia I, und Empis JMA. Isolation of bacterial strains capable of using lupanine, the predominant quinolizidine alkaloid in white lupin, as sole carbon and energy source. Journal of Industrial Microbiology. 1996;17:110-115. doi:10.1007/BF01570053

[21] Weissbach F. Beziehungen zwischen Ausgangsmaterial und Gärungsverlauf bei der Grünfuttersilierung [thesis]. Rostock: Universität Rostock; 1968.

[22] Münte H. Einsäuerungsversuche mit grünen Lupinen. Dissertation [thesis]. Kiel: Universität Kiel; 1931.

[23] Holzschuh W, Schmidt W. Silage. Herstellung, Fütterung. VEB Deutscher Landwirtschaftsverlag. 1963.

[24] Hopper DJ, Kaderbhai MA, Marriott SA, Young M, Rogozinski J. Cloning, sequencing and heterologous expression of the gene for lupanine hydroxylase, a quonocytochrome c from a Pseudomonas sp. Biochemical Journal. 2002;367:483-489. doi:10.1042/BJ20020729

[25] Hopper DJ, Rogozinski J, Toczko M. Lupanine hydroxylase, a quinocytochrome c from an alkaloid degrading Pseudomonas sp. Biochemical Journal. 1991;279:105-109. doi:10.1042/bj2790105

[26] Toczko M. Induced lupanine hydroxylase from Pseudomonas lupanini. Biochimica et Biophysica Acta (BBA)—Enzymology and Biological Oxidation. 1966;128:570-573. doi:10.1016/0926-6593(66)90018-X

[27] Toczko M, Brzeski W, Kakolewska-Baniuk A. Microbial degradation of lupanine. V. Identification of 17-hydroxylupanine. Bulletin of the Polish Academy of Sciences. 1963;Cl. II, 11: 161-164.

[28] Santana FC, Empis J. Bacterial removal of quinolizidine alkaloids from *Lupinus albus* flours. European Food Research and Technology. 2001;212:217-224. doi:10.1007/s002170000221

[29] Camacho L, Sierra C, Marcus D, Guzman E, Campos R, von Bäer D, Trugo L. Nutritional quality of lupine (*Lupinus albus* cv. Multolupa) as affected by lactic acid fermentation. International Journal of Food Microbiology. 1991;14:277-286. doi:10.1016/0168-1605(91)90119-A

[30] Gefrom A, Ott EM, Hoedtke S, Zeyner A. Ensiling legume seeds and the influence of conservation on contents of alkaloids, oligosaccharides, phytate-phosphorus and tannins. Züchtungskunde. 2013;85(2):154-168. ISSN:0044-5401

[31] Gefrom A, Ott EM, Hoedtke S, Zeyner A. Ensiling legume grains to reduce antinutritional factors; Effect of ensiling moist field bean (*Vicia faba*), pea (*Pisum sativum*) and lupine (Lupinus spp.) grains on the contents of alkaloids, oligosaccharides and tannins. Journal of Animal Physiology and Animal Nutrition. 2012. doi:10.1111/jpn.12024

[32] Martinez-Villaluenga C, Gulewicz P, Frias J, Vidal-Valverde C. Amino acid profile in fermented *Lupinus angustifolius* cv. Zapaton. Integrating legume biology for sustainable agriculture. In: Proceedings of the 6th European Conference on Grain legumes. 12-16 November 2007; Lisboa. Portugal. p. 181.

[33] Hackl W, Pieper B, Pieper R, Korn U, Zeyner A. Effects of ensiling cereal grains (barley, wheat, triticale and rye) on total and pre-caecal digestibility of proximate nutrients and amino acids in pigs. Journal of Animal Physiology and Animal Nutrition. 2010;94(6):729-735. doi:10.1111/j.1439-0396.2010.01032.x

[34] Frank A, Hackl W, Ott EM, Gefrom A, Zeyner A. Untersuchungen zum Einsatz von Lupinenkornsilage in der Ferkelaufzucht. In: Proceedings of the DLG Forum angewandte Forschung in der Rinder- und Schweinefütterung; 01-02 April 2009; Fulda. Germany. pp. 168-171.

[35] VDLUFA—Association of German Agricultural Analytic and Research Institutes e. V.: Handbook, Vol. III: Feedstuff Analysis, 3th ed. Darmstadt; 1997. 2190 p. ISBN: 978-3-941273-14-6

[36] Schmidt T, Krawielitzki K, Voihgt J, Gabel M. Methodical aspect of the in situ technique for estimating ruminal nutrient degradation: microbial contamination and lag time. Übersichten zur Tierernährung. 2005;33(2):87-100. ISSN:0303-6340

[37] Wink M, Meissner C, Witte L. Patterns of quinolizidine alkaloids in 56 species of the genus Lupinus. Phytochemistry. 1995;38:139-153. doi:10.1016/0031-9422(95)91890-D

[38] Hoedtke S, Zeyner A. Comparative evaluation of laboratory-scale silages using standard glass jar silages or vacuum-packed model silages. Journal of the Science of Food and Agriculture. 2011;91:841-849. doi:10.1002/jsfa.4255

[39] DLG (2006). Grobfutterbewertung, Teil B, DLG-Schlüssel zur Beurteilung der Gärqualität von Grünfuttersilage auf Basis der chemischen Untersuchung. Informationen 2/2006, DLG-Verlag, Frankfurt am Main. Available from: http://www.dlg.org/fileadmin/downloads/fachinfos/futtermittel/grobfutterbewertung_B.pdf [Accessed: 2016-09-27].

[40] Jansen G, Seddig S, Jürgens HU. Untersuchungen zum „Stärkegehalt" in Blauen Süßlupinen. 8. Gesellschaft für Pflanzenzüchtung-Tagung, Freising-Weihenstephan, 2006, 14.03.-16.03.2006, Vortrag Pflanzenzüchtung, 68, 73.

[41] White CL, Hanbury CD, Young P, Phillips N, Wiese SC, Milton JB, Davidson RH, Siddique KHM, Harris D. The nutritional value of *Lathyrus cicera* and *Lupinus angustifolius* grain for sheep. Animal Feed Science and Technology. 2002;99:45-64. doi:10.1016/S0377-8401(02)00035-4

[42] Wrigley C. The lupin the grain with no starch. Cereal Foods World. 2003;48:30-31.

[43] DLG (2014). Deutsche Landwirtschafts-Gesellschaft, DLG-Futterwerttabellen Schweine. DLG-Verlag, Frankfurt am Main.

[44] DLG (1997). Deutsche Landwirtschafts-Gesellschaft, DLG-Futterwerttabellen Schweine. DLG-Verlag, Frankfurt am Main.

[45] GfE (Ausschuss für Bedarfsnormen der Gesellschaft für Ernährungsphysiologie) Flachowsky G: Empfehlungen zur Energie- und Nährstoffversorgung von Schweinen. 1st ed. Band 10 von Energie- und Nährstoffbedarf landwirtschaftlicher Nutztiere. DLG Verlag, Frankfurt am Main; 2006. 247 p. doi:9783769006834

[46] WPSA (World's Poultry Science Association) Subcommittee Energie of the Working Group nr. 2 Nutrition of the Federation of Branches of the World's Poultry Science

Association, European Table of Energie Values for Poultry Feedstuffs, 3rd Edition. 1989, 104.

[47] Marquard R. III. Nutritive und antinutritive Inhaltsstoffe der Leguminosen. In: Schuster WH, Alkämper J, Marquard R, Stählin A, editors. Leguminosen zur Kornnutzung [Internet]. 2002. Available from: http://geb.uni-giessen.de/geb/volltexte/2000/320/ [Accessed: 2016-09-02].

[48] Römer P. Lupinen—Verwertung und Anbau [Internet]. 2007. Available from: http://lelf.brandenburg.de/sixcms/media.php/4055/lupine07.15564210.pdf [Accessed: 2016-09-02].

[49] Sujak A, Kotlarz A, Strobel W. Compositional and nutritional evaluation of several lupin seeds. Food Chemistry. 2006;98:711-719. doi:10.1016/j.foodchem.2005.06.036

[50] Pearson G, Carr JR. A comparison between meals prepared from the seeds of different varieties of lupin as protein supplements to barley-based diets for growing pigs. Animal Feed Science and Technology. 1977;2:49-58. doi:10.1016/0377-8401(77)90040-2

[51] Godfrey NW, Mercy AR, Emms Y, Payne HG. Tolerance of growing pigs to lupine alkaloids. Australian Journal of Experimental Agriculture and Animal Husbandry. 1985;25:791-795. doi:10.1071/EA9850791

[52] Bellof G, Sieghart S: Süßlupinen und Rapskuchen. Eiweißalternativen in der Schweinemast. Deutsche Geflügelwirtschaft und Schweineproduktion. 1996;5:45-49.

[53] Allen JG. Toxins and lupinosis. In: Gladstones JS, Atkins CA, Hamblin J, editors. Lupins as crop plants: biology, production and utilization. CAB International, Cambridge; 1998. pp. 411-435. doi:10.1023/A:1017590829523

[54] Petersen U, Schulz E. Examination about the suitability of faba beans (Vicia faba L. minor), sweet lupins (Lupinus luteus L.) and rapeseed meal (Brassica napus L. var napus) as protein source in pig fattening. 2. Landwirtsch. Forsch. 1978;31:269-280.